The Politics of Military Force

The Politics of Military Force examines the dynamics of discursive change that made participation in military operations possible against the background of German antimilitarist culture. Once considered a strict taboo, so-called out-of-area operations have now become widely considered by German policymakers to be without alternative. This book argues that understanding the facilitation of certain policies (in this case, military operations abroad and force transformation) requires focusing on processes of discursive change that make different policy options rational, appropriate, feasible, or even self-evident. Drawing on Essex School discourse theory, the book develops a theoretical framework for understanding the workings of discursive change and elaborates on how such change makes once unthinkable policy options not only acceptable but even without alternative. Based on a detailed discourse analysis of more than 25 years of German parliamentary debates, *The Politics of Military Force* provides an explanation for (1) the emergence of a new hegemonic discourse in German security policy after the end of the Cold War (discursive change), (2) the rearticulation of German antimilitarism in the process (ideational change / norm erosion), and (3) the resulting facilitation of military operations and force transformation (policy change). The book demonstrates the added value of a poststructuralist approach compared to the naive realism and linear conceptions of norm change so prominent in the study of German foreign policy and in the field of International Relations more generally.

Frank A. Stengel is a research fellow in the Research Group on International Political Sociology at Kiel University, Germany.

The Politics of Military Force

Antimilitarism, Ideational Change, and Post-Cold War German Security Discourse

Frank A. Stengel

University of Michigan Press
Ann Arbor

Copyright © 2020 by Frank A. Stengel

For questions or permissions, please contact um.press.perms@umich.edu

Published in the United States of America by the University of Michigan Press
Manufactured in the United States of America
Printed on acid-free paper
First published December 2020

A CIP catalog record for this book is available from the British Library.

Library of Congress Cataloging-in-Publication data has been applied for.

ISBN: 978-0-472-13221-8 (hardcover: alk. paper)
ISBN: 978-0-472-12731-3 (ebook)

Für Jana

Contents

Digital materials related to this title can be found on the Fulcrum platform
via the following citable URL: https://doi.org/10.3998/mpub.10154836

Foreword

Frank Stengel's study proceeds from a fairly straightforward empirical puzzle: why, despite decades of strong public commitments opposing the use of military force, has Germany altered its stance such that participation in out-of-area military operations has become an everyday part of German security practices? Existing explanations for the Federal Republic of Germany's previous anti-military stance make reference to deeply-embedded norms rejecting the use of force, the rapid shift in policy raises serious problems for any such account. And explanations for the participation of German military forces in a wide range of novel deployments that rely on supposedly obvious facts about the changed security environment run into immense difficulties when confronted with the ambiguities of that environment—and the plethora of interpretations and arguments about just what kind of response is called for.

The solution to these explanatory deficiencies, Stengel proposes, is to take more seriously the process by which state actions are justified and legitimated. To bring this process into view, Stengel deploys the tools of critical discourse analysis, specifically its Essex School variant, and concentrates on how situated policymakers and politicians take advantage of gaps and mismatches between elements of German security discourse to remake a stance opposing the use of military force into a stance that supports that use. In particular, Stengel notes, the mutation of the notion that military force is a tool of last resort from a component of an argument *against* out-of-area troop deployments to a component of an argument *for* such deployments helps to explain why the conditions of possibility for German security practices changed. This is not a change of overall "ideas," because the same notion— military force is a tool of last resort—is present in *both* arguments. Instead, the overall arrangement of discursive elements has changed, and that alters the meaning of specific notions. Something similar, Stengel argues, is at work in the reinterpretations of the German past and the drawing of different "lessons of history" from the experience of Nazism; where once this was a cautionary tale about the dangers of militarism, now it is a cautionary tale about the kinds of evils that can *only* be opposed by military force.

Stengel's argument is considerably more than an explanation of one country's changing foreign policy, however. Against both interest- and idea-based accounts of social action, especially social action undertaken by artificial persons like states, Stengel illustrates how much more we learn about international affairs when we focus our attention on processes of articulation and concrete strategies of meaning-making. Against more or less determinist accounts of political change, Stengel recovers contingency and agency without sliding into an unfettered indeterminism. Both of these broad theoretical moves have implications for how we think about international affairs in general, since they point to the need to take more seriously the specifics of how actions become possible in particular contexts, and how different discursive configurations give rise to different outcomes. Indeed, an extension of Stengel's analysis to other otherwise-puzzling cases is entirely reasonable, and in Stengel's book, scholars will have an outstanding example of just what such an account looks like.

I would also be remiss if I did not highlight the relevance of Stengel's argument for contemporary discussions and debates about the future of international order. A demilitarized Germany was a core component of the "Western" geopolitical settlement that followed the Second World War, and along with the transatlantic alliance instantiated in NATO, formed the institutional architecture that tacitly enframed world politics in the Euro-American "core" of the world-system for decades. If we did not pay attention to the way that a demilitarized Germany became a militarily active Germany, we would likely mis-estimate the durability of that "Western" geopolitical settlement. The changes Stengel points to did not take the form of a complete break with past political coalitions and positions, but are best understood as a contingent evolution of how a variety of discursive elements were configured. While that does not initially present an optimistic view of the stability of our global institutions, it also calls for a more responsible exercise of political agency in defense of those institutions. In that way, Stengel's reflections on the past has clear lessons for us in the present, as a good piece of critical social science should.

Patrick Thaddeus Jackson
Series Editor, Configurations

Preface and Acknowledgments

This book is about the intersection of changes in discourse and foreign policy, in the context of security policy. It grew out of my PhD dissertation, which I began at the Bremen International Graduate School of Social Sciences (BIGSSS) at the University of Bremen and finished at Kiel University. My research in Bremen was funded by a full scholarship financed by the German Excellence Initiative, and the project received additional funding by a research scholarship from the German Institute of Global and Area Studies (GIGA), as well as a completion grant by the University of Bremen. I finished the final manuscript at the School of Advanced International Studies (SAIS) at Johns Hopkins University, while on a postdoctoral fellowship financed by the German Academic Exchange Service.

Over the years, I profited immensely from comments by many people on various chapter drafts and conference papers. First and foremost in this context, I thank my PhD supervisors at BIGSSS—Martin Nonhoff, Rainer Baumann, Thomas Diez, and Dirk Nabers—for their invaluable comments, patience, and support over the years; they went well beyond what can be expected. The manuscript profited from discussions at BIGSSS with other fellows, most notably Jesse Crane-Seeber and Anup Sam Ninan, as well as the members of what we called our "Discourse Self-Help Group," Dominika Biegoń and Linda Monsees. I presented parts of the project at a number of conferences and workshops, including doctoral colloquia at GIGA in Hamburg, the Institute for Intercultural and International Studies at the University of Bremen, and the University of Marburg; the methodology workshop at the 2009 Interpretive Policy Analysis Conference at the University of Kassel; a research colloquium on German foreign policy at the Université Jean Monnet in Saint-Étienne, 12–13 November 2009; the annual meetings of the International Studies Association in Montréal in 2010 and in Atlanta in 2016 (a shout-out goes to my ISA "buddies" David MacDonald and Robert Pat-man); and a workshop on discourse analysis at BIGSSS in 2013. I thank the participants in these various events, particularly Ken McDonagh, Lene Hansen, and Richard Samuels, for helpful comments.

This project profited equally from individual discussions with Marcus Beiner, Jana Jarren, Peter Mayer, Ryoma Sakaeda, David Shim, Wilfried Stengel, and Bernhard Zangl, at various stages of the research. Christoph Weller has been extremely supportive; without his advice and support, the project might have died before it started. The book manuscript itself was finished at Kiel University, where, in addition to Dirk Nabers, my colleagues Merve Genç, Malte Kayßer, and Jan Zeemann helpfully read and commented on individual chapters. I thank Merve Genç for an excellent job creating some of the figures in this study and primping up the others, and I thank her and Friederike Bartels for going over the final manuscript with a fine-tooth comb. Nadine Klopf provided much-needed help with the index. At SAIS, Jason Moyer helped me get my figures into a printable format. I am grateful to Nick Smith not only for early discussions about discourse theory but for indispensable advice, years later, on turning a thesis into a book. In addition to Nick, I thank David MacDonald for providing very helpful advice in that regard.

That this book has been published by the University of Michigan Press is very much due to Patrick T. Jackson and Elizabeth Demers, who, luckily for me, took an interest in the project from the start and have been extremely supportive along the way. I am grateful to Patrick for providing multiple rounds of comments that were instrumental in revising the manuscript. Through their highly professional handling of the overall process at the press, Elizabeth, Danielle Coty, and Mary Hashman made things much easier for me than they could have been. I would like to thank Jill Butler Wilson for doing a great job whipping my sometimes slightly Germanic English into a readable form.

I offer my sincere gratitude to two anonymous reviewers for two rounds of extremely constructive reviews, first on my book proposal and sample chapters, then on the full manuscript. I do not exaggerate a bit when I say that both reviewers were the exact opposite of the infamous Reviewer 2; they were thorough, rigorous, extremely well informed, impressively attentive to detail, and committed to improving the manuscript rather than merely judging it. Because of their insistent (but friendly) probing into the manuscript's weak spots, the book has turned out much better than it would otherwise be.

I thank my parents, brothers, in-laws, and son, Fiete, as well as a handful of close friends, all of whom suffered, to different extents, under what sometimes seemed like a never-ending project. Above all, I am grateful to my wife, Jana Jarren, for her patience, support, encouragement, and faith in my ability to write a book. Without Jana, this book, dedicated to her, would have never been written.

Abbreviations

AA	Federal Foreign Office (Auswärtiges Amt)
BMZ	Federal Ministry for Economic Cooperation and Development (Bundesministerium für wirtschaftliche Zusammenarbeit und Entwicklung)
BGBl	Federal Law Gazette (Bundesgesetzblatt)
BMVg	Federal Ministry of Defense (Bundesministerium der Verteidigung)
BVerfG	(German) Federal Constitutional Court (Bundesverfassungsgericht)
BVerfGE	Decisions of the German Federal Constitutional Court (Entscheidungen des Bundesverfassungsgerichts)
CDU	Christian Democratic Union of Germany (Christlich Demokratische Union Deutschlands)
CSU	Christian Social Union in Bavaria (Christlich Soziale Union in Bayern)
EU	European Union
FDP	Free Democratic Party (Freie Demokratische Partei)
FRG	Federal Republic of Germany (Bundesrepublik Deutschland)
GDR	German Democratic Republic (Deutsche Demokratische Republik)
GG	Grundgesetz (Basic Law of the Federal Republic of Germany)
GOBT	Geschäftsordnung des Deutschen Bundestages (Rules of Procedure of the German Bundestag)
ISAF	International Security Assistance Force in Afghanistan and Uzbekistan
IOs	international organizations
IR	International Relations

MP	member of parliament (Bundestag)
NATO	North Atlantic Treaty Organization
OECD	Organization for Economic Co-operation and Development
OEF	Operation Enduring Freedom
OEF-A	Operation Enduring Freedom in Afghanistan
PDS	Party of Democratic Socialism (Partei des Demokratischen Sozialismus; preceded by the SED, succeeded by Die Linke)
QRF	Quick Reaction Force
RAF	Red Army Faction (Rote Armee Fraktion)
SED	Socialist Unity Party (Sozialistische Einheitspartei Deutschlands)
SPD	Social Democratic Party of Germany (Sozialdemokratische Partei Deutschlands)
UK	United Kingdom
UN	United Nations
UNOSOM II	United Nations Operation in Somalia
UNPROFOR	United Nations Protection Force in Bosnia
US	United States
WEU	Western European Union
WMD	Weapons of mass destruction

Introduction

Today, the Bundeswehr is an army on operation
(interjection by Die Linke: "Boo!");
national defense takes place also at the Hindu Kush.

GERMAN DEFENSE MINISTER PETER STRUCK[1]

This book analyzes changes in the German security discourse after unification, focusing specifically on the emergence of "networked security" (*vernetzte Sicherheit*) as the overarching framework for post–Cold War German security policy. In doing so, it follows two main avenues. First, for understanding processes of discursive change, it proposes a theoretical framework based on the poststructuralist discourse theory of the so-called Essex School (Glynos and Howarth 2007; Howarth et al. 2000; Laclau and Mouffe 2001). Although primarily located within poststructuralist theory,[2] this study demonstrates that understanding discursive change is highly relevant to a much broader range of theoretical questions, including norm dynamics, the transformation of taboos in international relations, grand strategic change, the "identity-security nexus" (e.g., Innes 2010), and the domestic or international legitimation/making-possible of certain (potentially controversial) policies (Doty 1993; Nuñez-Mietz 2018; Wajner 2019; Weldes and Saco 1996),

---

1. Speech at the German Bundestag, 16th legislative period, 2nd session, 8 November 2005: 43. Struck made this statement for the first time at a press conference in December 2002 (von Bredow 2015: 153). The sources for subsequent quotes from parliamentary protocols are provided in the form of in-text short citations according to the following template: name of the speaker, legislative period/session number, date: page. All translations from the German language in this book are, if not otherwise indicated, the author's, including parliamentary protocols and the German-language academic literature.

2. *Poststructuralism* is a problematic and highly contested term (see Angermüller 2015). For practical reasons, I employ it here to refer to the approach used in this study.

most notably the threat and use of military force. Second, the book inter-
venes in the ongoing scholarly debates about German security policy, particularly German participation in military operations outside the North
Atlantic Treaty Organization (NATO) area. Despite a supposedly widespread
and deeply engrained antimilitarist culture (e.g., Berger 1998), German participation in these so-called out-of-area operations has increased significantly since the end of the Cold War. Moreover, such operations—once perceived as "completely unthinkable," in the words of German Chancellor
Angela Merkel (17/37, 22 April 2010: 3478)[3]—have become widely accepted
among German policymakers as a normal (if unpopular) element of German
security policy. The "out-of-area debate" (Dalgaard-Nielsen 2006; Longhurst
2004) has given way, it seems, to an out-of-area consensus.[4]

This book sets out to provide an answer to the puzzle of how German
policymakers have widely taken out-of-area operations for granted as a social
practice. Assuming a discourse theoretical perspective, the book argues that
whether policies are considered appropriate or inappropriate, rational or
irrational, or moral or immoral depends on the discursive order (the established, dominant discourse) that organizes a certain field of human activity
at a certain point in time.[5] Starting from this basic argument, shared by (the
partially overlapping fields of) discourse theory, poststructuralism in International Relations (IR), and (parts of) International Political Sociology (IPS)
and critical constructivism alike,[6] the book proposes that understanding
how once-unthinkable policies are made possible (i.e., how taboos erode)
requires that we turn our attention to dynamics of discursive change. Thus,
that out-of-area operations have become not only acceptable but considered
a self-evident requirement of a post–Cold War world can only be understood
in the broader context of changes in the German security discourse. How
"reality" (including a state's security environment) is understood, who or

---

3. On taboos in IR, see Tannenwald 1999, as well as the contributions to the fourth
issue of *Review of International Studies* 36 (2010).

4. This consensus is discussed in the literature as Germany's "normalization"
(Crawford 2010; Gordon 1994; Karp 2009; critical, Kundnani 2012) or its "coming-
of-age" (Brockmeier and Rotmann 2018: 20).

5. In that sense, discourse functions in a similar way to how strategic culture is
sometimes conceptualized (e.g., Biehl et al. 2013; Longhurst 2004). I prefer discourse
because the "beliefs, norms and ideas" (Biehl et al. 2013: 11) of which culture consists
can only be observed empirically once they are articulated in discourse.

6. See, e.g., Ashley and Walker 1990; Behnke 2013; Campbell 1998; Der Derian
and Shapiro 1989; Doty 1993; Hansen 2006; Nabers 2015; Nabers and Stengel 2019a;
Weldes and Saco 1996; Weldes et al. 1999; Zehfuss 2002.

what is considered a threat, what means are appropriate to solve certain policy problems, and how specific norms and values (e.g., antimilitarism) should be understood in the context of security policy is produced, regulated by, and transformed in the security discourse. In short, security discourses are primarily concerned with what is usually referred to as grand strategy, broadly understood as "a state's theory about how it can best 'cause' security for itself" (Posen 1986: 13; see Krebs 2018).

Building on these arguments, this book examines dynamics of discursive change, with the ultimate aim to add to our understanding of (foreign) policy change. After developing a theoretical framework to analyze what makes some discourses more effective than others, it provides an explanation, based on a comprehensive discourse analysis of more than 25 years of German parliamentary debates, for the transformation of the German security discourse since the Cold War. Moreover, the book traces how military operations have been articulated differently in the Cold War and the current discursive order, making them unthinkable (a taboo) in one case and without alternative in the other.

WHY INTERVENTIONISM IS NOT SELF-EVIDENT: PROBLEMATIZING THE GERMAN OUT-OF-AREA CONSENSUS

The starting point of this book is the curious expansion of involvement abroad by the German armed forces, which presents a puzzle for existing theoretical accounts. Over the past 30 years, the Bundeswehr—the official title of the German armed forces—has undergone nothing less than a "dramatic transformation" (Enskat and Masala 2015: 365) from a "non-interventionist, conscription-based territorial defense force" (Sarotte 2001: 12) to an "army on operation" (e.g., Jung, 16/227, 18 June 2009: 25169).[7] Once confined to territorial defense within NATO, the role of the Bundeswehr has been gradually expanded to include conflict prevention, crisis management, and counterterrorism, as core functions.[8]

---

7. The original German term *Armee im Einsatz* has been translated inconsistently in official documents as "army on operations" (Federal Ministry of Defence 2003: 18), "expeditionary force" (Federal Ministry of Defence 2006: 6), and, in the 2005 coalition agreement between CDU, CSU, and SPD, "operational army" (CDU et al. 2005: 126). Often, this is simply called the "new Bundeswehr" (BMVg 2012a: 9).

8. While many of the Bundeswehr's current tasks, including conflict prevention, can be traced back to the 1994 white paper on security (Federal Ministry of Defence

Importantly, while individual missions continue to be controversially debated,[9] the general practice of out-of-area operations has become a largely uncontroversial matter among members of the Bundestag, the German parliament (with the notable exception of the left-wing party Die Linke). Today, that the Bundeswehr should participate in military operations around the globe has become, for the majority of German policymakers, a self-evident fact of life (Enskat and Masala 2015; von Bredow 2015; von Krause 2013, 2015). Moreover, force transformation with the explicit aim of making the Bundeswehr fit for its changed "operational reality" (Struck, 15/97, 11 March 2004: 8601) has become a constant feature, bringing about numerous legal reforms and quite material consequences in new arms procurement plans. In short, out-of-area operations have become a social practice, understood as "the ongoing, routinized forms of human and societal reproduction" that are mostly taken for granted, without any "strong notion of self-conscious reflexivity" (Glynos and Howarth 2007: 104).

This general consensus, normal though it might seem in a time in which most "Western" countries pursue interventionist security policies,[10] is actually highly remarkable, for at least two reasons. First, the Federal Republic of Germany (FRG) is commonly considered to be an ideal-type "civilian power" (Maull 1990, 2018) whose foreign policy is marked by a preference for multilateralism, Western integration, and, above all, antimilitarism (Baumann 2011: 468)—an "extraordinary reluctance to become actively involved in

---

1994), the paper, published before the Federal Constitutional Court's 1994 out-of-area decision, only cautiously advocates a more active international role (Martinsen 2010). The paper quite clearly places emphasis on national defense as the Bundeswehr's main function, explicitly arguing that the maintenance of a "protective function must not be affected by changes in the security situation," because armed forces were intended mainly as an insurance "against the imponderables of the future" (Federal Ministry of Defence 1994: para. 302). In contrast, the 2006 white paper lists conflict prevention, crisis management, and counterterrorism first when discussing the Bundeswehr's functions, even prior to the "support of allies" and the "protection of German territory and its citizens" (Federal Ministry of Defence 2006: 9). While the change is gradual, it is still substantial.

9. That not each and every mission is surefire business for Germany is clearly demonstrated by the country's 2011 abstention from United Nations (UN) Security Council Resolution 1973 authorizing the Libya intervention (see Brockmeier 2013; Meiers 2012; Miskimmon 2012).

10. I intend the scare quotes to indicate that the concept of "the West" itself is discursively produced (see Behnke 2013; Hall 1992; Hellmann and Herborth 2016; Klein 1990; on "Western" interventions, Dillon and Reid 2009; Kühn 2013; Orford 1999; Sabaratnam 2018; Zehfuss 2018).

international military security affairs" (Berger 1998: 1).[11] This argument is primarily made by conventional constructivist scholars who point to the importance of various relatively stable "ideational variables" (Malici 2006: 37)—norms, values, roles, identities, political cultures, and so on[12]—that, internalized by policymakers and the general public alike, influence standards of appropriateness. As a consequence, constructivist scholarship would have led us to expect antimilitarist culture to have a constraining effect and to function as a formidable obstacle to military involvement abroad (Longhurst 2004: 131; Crawford and Olsen 2017).[13] This expectation should apply especially to violent missions like that of the International Security Assistance Force (ISAF) in Afghanistan (see Noetzel 2011). The expansion of out-of-area operations is rendered even more puzzling because antimilitarism does not seem to have significantly weakened among the general public.[14] This point is of interest not only to researchers concerned with German foreign policy but, equally, to students of Japanese foreign policy. Like Germany, post-1945 Japanese foreign policy has traditionally adhered to a strict antimilitarism but recently began a process of "normalization" (e.g., Hughes 2009; Stengel 2007; recently, Gustafsson et al. 2018), despite strong, if declining, public opposition (Hagström and Isaksson 2019). As in the German case, scholars struggle to provide convincing explanations for this puzzle (but see, recently, Gustafsson et al. 2019; Hagström and Hanssen 2016; Hagström and Isaksson 2019).

Second, although policymakers themselves usually attribute expansion of out-of-area operations to the pressures of a changed security environment, a closer look renders this argument unconvincing. After the end of the Cold War, we are told, the world is marked no longer by traditional threats

---

11. See also, recently, Crossley-Frolick 2013: 43; Daase and Junk 2012; Hilpert 2014; Leithner 2009. Others have called this the *Kultur der Zurückhaltung*, translated as "culture of restraint" (Longhurst 2004: 130) or "culture of reticence" (Crawford 2010: 181; Malici 2006).

12. Among studies drawing on ideational variables, political, strategic, and/or security cultures are the most prominent (see Berger 1998; Daase and Junk 2012; Duffield 1998; Giegerich and von Hlatky 2019; Hilpert 2014; Junk and Daase 2013; Lantis 2002a; Malici 2006), but scholars have also explained policy by drawing on norms and values (Baumann 2001; Boekle et al. 2001), roles (Koenig 2020; Maull 2018), identities (Banchoff 1999; Risse 2007), and even Bourdieu's (1984) concept of habitus (see Bjola and Kornprobst 2007).

13. See, e.g., Berger 1998; Bjola and Kornprobst 2007; Maull 2000. For a more nuanced take, see Junk and Daase 2013.

14. For data, see Gravelle et al. 2017; Körber-Stiftung 2014; Mader 2017; Schoen 2010.

but by "new" globalized threats like terrorism, state failure, or the proliferation of weapons of mass destruction that, if not countered early and at their source of origin, will eventually reach Germany as well. For example, in a much-discussed speech in 2014, federal president Joachim Gauck claimed that since Germany could not "hope to be spared from the world's conflicts," the country "should not [*darf nicht*] say 'no' on principle" to military operations (Gauck 2014: 5). The implicit logic behind this claim is one political scientists call (neo)functionalist: policy measures are responses to objective problems.[15]

There is one major problem with this line of reasoning: the claim that new threats like terrorism, intrastate conflict, or state failure demand out-of-area operations does not sit easily with the literature concerned with the effectiveness of military operations. Although systematic evidence is hard to come by, the recent literature suggests that military operations are of very limited, if any, utility to counter new threats like terrorism, intrastate conflict, or state failure, let alone climate change or mass migration.[16] Even with respect to traditional peacekeeping operations, which take place after a cease-fire or peace agreement has been reached and which are commonly regarded as the most successful type of interventions (Fortna 2004; Fortna and Howard 2008; Gromes 2012), success seems to vary with different factors, including whether it is a UN mission (Nilsson 2008), the mission's mandate (Salvatore and Ruggieri 2017), whether it includes a civilian component (Hoeffler 2014), and whether it is a so-called robust (Bellamy and Hunt 2015: 1280) or militarized (Sloan 2011) operation.[17] In addition, because military operations at least entail the possibility of violence, they risk creating unintended negative consequences like injuring civilians and/or provoking resistance (Condra and Shapiro 2012; Hughes 2015: 106; Johnson 2004a;

---

15. Similar arguments are prominent in the debate about Japanese security policy, although the rise of China and North Korean aggression are commonly seen as the most important factors causing foreign policy change in Japan (e.g., Hughes 2004; Samuels 2007). As in the German case, this interpretation is not self-evident (see Gustafsson et al. 2019). The same can be said about arguments that pressures from external problems require Japanese participation in peace operations (Stengel 2008).

16. See Bellamy 2015; Downes and Monten 2013; Gilligan and Sergenti 2008; Grimm 2008; Gromes and Dembinsky 2013; Mac Ginty 2012; Regan and Meachum 2014.

17. Even peacebuilding, which uses primarily civilian instruments, has been subject to sustained criticism, raising doubt concerning the effectiveness of external interventions more generally (Autesserre 2017; Duffield 2007; Goetze 2017; Richmond 2011; Sabaratnam 2018).

Paris 2014).[18] Following some of these studies' recommendations would mean less, not more, military activity abroad. In summation, although assessing the effectiveness of military operations is an arduously complicated task fraught with methodological problems, it is relatively safe to say that the available research at least does not support the claim that new threats demand military interventions. Of course, out-of-area operations equally serve the purpose of demonstrating solidarity within NATO and the European Union (EU) (Kaim 2007), but that purpose only shifts the question from Germany to its partners, because, from a functionalist perspective, it is no less clear why the United States (US) would rely on largely unsuitable policy instruments.

Given these limitations, it is, as Enskat and Masala aptly summarize, "not self-evident but to the highest degree remarkable" that out-of-area operations have "become almost a thing of course" (2015: 373). This book takes this puzzle as a starting point, arguing that the German out-of-area consensus can only be understood within the context of large-scale discursive change. Put simply, the book argues that the changing view on military operations is the result of the demise of one (the Cold War) security order and its replacement by another.

COMPETING EXPLANATIONS

Previous attempts at explaining Germany's move into out-of-area operations fall within three (and a half) broad categories.[19] The first group of studies explains out-of-area operations (as do policymakers themselves) as an adaptation to changed circumstances. They include, most notably, neoclassical realist and theoretically eclectic, policy-oriented studies. For example,

---

18. This concern is in addition to any ethical questions (see Baron 2010; Dill and Shue 2012; McMahan 2009; Rudolf 2014; Zehfuss 2018). Indeed, recent years have seen pacifism and nonviolence return as serious topics in IR (Frazer and Hutchings 2014; Hutchings 2018; R. Jackson 2019).

19. I limit my discussion here to explanations of the more general shift toward the out-of-area consensus, not of individual policy decisions. The literature on German security policy is wider than the discussion here suggests, including, at least during recent years, insightful studies from the perspective of Foreign Policy Analysis, which explain individual operations (see Brummer 2011, 2012, 2013; Brummer and Oppermann 2019) as well as focus on the effect of party ideologies and contestation on variation within the broad limits set by ideational factors (or discourses) (Hofmann 2019; Wagner et al. 2018).

criticizing constructivist studies, Dyson argues that Germany was simply "acting according to the material forces of the international system, rather than subjective norms and ideas rooted in German 'security culture'" (2011: 559; also 2019). Similarly, Glatz et al. (2018) state that out-of-area operations are "necessary," and Buras and Longhurst claim,

> The *international situation after the end of the Cold War*, Germany's acquisition of full sovereignty coupled with demands from allies and partners to take up a greater responsibility for security and stability in the world *necessitated a certain adjustment* of Bonn/Berlin's foreign and security policy. (2004: 226, italics added)

This argument, perhaps convincing at first glance, is problematic because it (implicitly or explicitly) takes reality as objectively given. This assumption stands in stark contrast to a diverse group of studies in IR, the social sciences, and philosophy that highlight that reality precisely cannot be taken for granted.[20] Ignoring their arguments is problematic for two reasons. First, taking one particular construction of reality as an objective representation of reality brackets a large portion of the politics involved in decision-making on matters of foreign policy. As a consequence, it offers a partial explanation at best. Second, such research actually reproduces one specific representation of reality and contributes to its enduring influence, including potential negative unintended consequences (Cox 1981; Dillon 1996; Smith 2004).

Despite the obvious limitations of this argument, it is shared even by some constructivist studies. For example, Leithner (2009: 9) explains discursive change as a result of "pressure from the new international environment" (similarly, Maull 2006). From a constructivist perspective, this argument is nothing less than self-defeating. Given that constructivists generally hold "the view that the material world does not come classified, and that, therefore, the objects of our knowledge are not independent of our interpretations and our language" (Adler 2002: 95), they should be among the first to point to the "social construction of reality" instead of taking it for granted (Berger and Luckmann 1967). From a constructivist point of view, claiming

---

20. This argument has been made in a broad range of studies based on very different theoretical positions across the social sciences: see, e.g., Ashley 1987; Behnke 2013; Berger and Luckmann 1967; Campbell 1998; Dillon 1996; George 1994; Hajer 2005; Hansen 2006; Houghton 1996; Jervis 1976; 2006; Mintz and Redd 2003; Sylvan 1998; Weldes 1999; Wendt 1995; Winch 1990.

that reality "demands" anything amounts to an ad hoc abandonment of one's (meta)theoretical framework as if it was a sweater, not a skin (Marsh and Furlong 2002), which is why Eberle is spot-on when he describes these studies as (only) "soft-constructivist" (2019: 4). Needless to say, this abandonment poses serious problems in terms of theoretical coherence (Guzzini 2000; P. T. Jackson 2010). Even if we were to gloss over these obvious inconsistencies (which we should not), the reference to an external reality means that constructivism itself has nothing to add and has to fall back on the theoretical competition.[21] A similar criticism applies to studies that invoke campaign tactics to explain the 2003 Iraq War (Dalgaard-Nielsen 2003: 100–101; Risse 2007: 59), which is to draw on rationalist explanatory factors rather than delivering a constructivist analysis.

The second group of explanations seeks to account for out-of-area operations from within constructivism. Most notably, Berger (2002) has argued that German policy change is the product of norm change, weakening antimilitarism in favor of multilateralism.[22] More than 15 years after Berger originally made the argument, its shortcomings are readily apparent. Thus, while participation in Operation Enduring Freedom in 2001 (the Afghanistan invasion) is compatible with the argument, German refusals to participate in the Iraq War in 2003 and the intervention in Libya in 2011 suggest that Germany has reverted to antimilitarism (Nonhoff and Stengel 2014). As has been pointed out by Baumann (2006), such a linear conception of norm change, which underpins much of conventional constructivist research (in both German foreign policy research and IR more generally: see Puetter and Wiener 2007), is not fully convincing, because it means that norms either are stable and constraining or change almost randomly back and forth (see also the critique in Flockhart 2016).

A third group of studies (the half category mentioned above) engage the issue but fall short of offering an explanation. These studies come in basically two variants. The first variant argues that change has actually been only incremental and moderate, still being compatible with German antimilita-

---

21. At this point, some observers will feel reminded of Legro and Moravcsik's (1999, 2000) highly similar and devastating critique of realism. Given the extent to which factors other than the international system were accountable for explaining policy outcomes in realist explanations of foreign policy, Legro and Moravcsik questioned whether anybody was still (or ever was) a realist.

22. Harnisch (2001) makes a similar argument with respect to socialization; more recently, Koenig (2020) has argued that Germany has undergone a role adaptation, placing more emphasis on multilateralism at the expense of military restraint.

rist culture and/or its civilian power role.[23] Aside from the fact that the extent of change is always a matter of interpretation (Baumann 2006; Hellmann 2009b), these studies simply state that constructivism passed the test and lives to fight another day. The second variant declares out-of-area operations outside of constructivism's jurisdiction altogether (Risse 2007: 59). Obviously, that approach provides no more insight.

Finally, a small, slowly growing body of research applies insights from critical IR and social theory to the study of German foreign policy, including discourse.[24] These studies significantly broaden our understanding of German foreign policy. Nevertheless, research that uses discourse as a main analytical concept remains slim (Crossley-Frolick 2017), and the few such studies focus on topics other than military operations or only on individual missions, concentrate on questions other than explaining large-scale discursive change, approach the issue from a different theoretical vantage point, and/or focus primarily on popular culture. With a macrolevel perspective on the changing German security discourse, the present study complements those previous ones.

ARGUMENT: POLICY CHANGE AND DISCURSIVE TRANSFORMATION

Overall, the widespread acceptance of out-of-area operations remains, to put it in more conventional political science terminology, a puzzle in need of explanation (Day and Koivu 2019; King et al. 1994: 15). This book shows that the common-sense assumption that military operations are essential is the result not so much of (what is commonly said to be) factual necessities but of a particular, contingent representation of reality within the German security discourse, rather than reality itself (which from a poststructuralist point of view is unintelligible anyway). To understand the establishment of military operations as a social practice (policy change), one needs to take a step

---

23. The basic argument that Germany remains a civilian power continues to be widespread: see Bjola and Kornprobst 2007; Koenig 2020; Malici 2006; Maull 2006, 2018; Müller and Wolff 2011; Risse 2007; critical, Hellmann 2002, 2007, 2011, 2016b.

24. See Bach 1999; Baumann 2006; Behnke 2012; Eberle 2019; Engelkamp and Offermann 2012; Geis and Pfeifer 2017; Hellmann 1999, 2007; Nonhoff and Stengel 2014; Roos 2012; Schoenes 2011; Shim and Stengel 2017; Spencer 2014; Stark Urrestarazu 2015; Stengel 2019b; Zehfuss 2002, 2007; Ziai 2010.

back and examine the much larger changes in the German security discourse as a whole (discursive change). In the, roughly, past 30 years, the Cold War German security order has become replaced by what I call, for convenience's sake, the "discourse of networked security." According to the new German grand strategy that this discourse produces, the old threat of the Soviet Union has been replaced by new threats like terrorism, mass migration, and environmental problems. Since these threats are globalized, they cannot be deterred but require a networked or comprehensive security policy that tackles them early on and at the place of their origin,[25] while combining the military and civilian instruments of different actors into a unified approach. In short, a networked security has to be both preventive and (in a broad sense) interventionist. Within this discourse, out-of-area operations are rearticulated in two important respects. First, military operations are constructed as indispensable within a broader whole-of-government strategy. Second, there is a transformation in the relationship between military operations as an instrument, on one hand, and peace and security as policy goals, on the other: once seen as contrary to peace and security, military operations (including the use of military force) have become accepted as a means to achieve peace and security.

To understand how the changing articulation of the military within the German security discourse has been made possible, this book traces and provides an explanation for the "hegemonization" of the discourse of networked security (Nabers 2015: 110; Norval 2004: 145). The term *hegemonization* here refers to the process by which a particular discourse manages to assert itself in discursive struggles, successfully establishing itself as "a valid and/or dominant world description" (Nonhoff 2019: 63). In this context, three aspects especially contribute to a particular hegemonic project's chance of success: (1) the construction of a broad range of social demands as equivalent (as going hand in hand), (2) the articulation of an antagonistic frontier between the Self and a radically threatening Other (that blocks the Self's very identity), and (3) the representation of the totality of equivalent demands by one particular demand (an empty signifier). Put simply, incorporating a broad range of demands increases the chance of gaining sufficient supporters to become hegemonic, and the identification of a clearly discernible root of all evil to be overcome galvanizes het-

---

25. This policy is also sometimes referred to in the literature as "extended security" (Junk and Daase 2013: 139).

erogeneous demands into a single project, which is further supported by the provision of a common symbol around which subjects can rally (Laclau 2005a; Laclau and Mouffe 2001). In addition to these three factors, a project has to be credible when held against the set of sedimented discursive practices that make up the normative framework of a given society (Laclau 1990a).

Here in particular, I argue, a discourse theoretical approach can benefit from engaging with arguments from feminist (e.g., Hooper 2001; Peterson and Runyan 1993; Sjoberg and Tickner 2013; Tickner 1988; Wibben 2018; Zalewski and Parpart 2008) and decolonial and postcolonial approaches as well as with arguments from critical geopolitics in geography (Dalby 1994; Ó Tuathail 1994, 1996).[26] The legitimation of military operations and, in particular, of the use of force is a prime example that some phenomena cannot be fully understood without taking feminist and postcolonial arguments into consideration. Indeed, a significant reason arguments for interventions (broadly understood) appear convincing is because, at the risk of oversimplification, they draw on established gendered representations (e.g., Young's "logic of masculinist protection"; see Young 2003) and civilizationist representations of a "modern" West and a "traditional" non-Western Other, which, in turn, are linked to older constructions of colonizer and colonized (Chakrabarty 2000; Cockburn 2010; Eichler 2014; Masters 2009; Muppidi 2012; Peterson 2010; Shepherd 2006). By drawing on that body of research, this book tries to respond to the criticism that "nonfeminist" research (including critical IR) does not sufficiently engage with feminist or postcolonial arguments (Åhäll 2018: 2; Chowdhry and Nair 2004a; Steans 2003; Tickner 1997; Wibben 2020; Zalewski 2019). To avoid silencing the importance of gender and Eurocentrism and the continued relevance of colonial discourses for the legitimation of interventions (and severely limiting explanatory power in the process), this book follows the proposal by Ann Towns (2019) to weave feminist and postcolonial arguments into the analysis. Having said that, readers should be aware that the gender and postcolonial analyses here remain limited in the sense that the book is primarily informed by the Essex

---

26. Decolonial and postcolonial perspectives comprise a very heterogeneous group. In the following discussion, I am using postcolonialism as a shorthand to refer to this body of research, but that should not provoke the misconception that this group of approaches is monolithic (on postcolonialism in IR, see Barkawi and Laffey 2006; Chandler 2013; Chowdhry and Nair 2004a; Darby 2009; Dunn 2003; Grovogui 2010; Inayatullah 2014; Vucetic and Persaud 2018).

School and, as a consequence, unavoidably falls short of fully realizing the "radical potential" of these perspectives.[27]

This book shows how the discourse of networked security prevailed. First, it transcended the confines of the security discourse more narrowly understood, by incorporating a number of different social demands, ranging from the security of Germany and its allies, to humanitarian concerns, to environmental protection. It included not only demands previously considered disparate but even some that were formerly seen to be contradictory—most notably, demands for out-of-area operations and civilian conflict prevention. Networked security thereby united previously opposing demands into a single hegemonic project. Second, the project clearly identified the source of enduring problems after the end of the Cold War, which, against original expectations, had not brought about world peace. This alleged root of all evil was the so-called new threats, which were articulated as a danger not just to Germany but to the entire international community, blocking it from fully establishing itself as a stable, democratic, peaceful, and perfectly secure entity. Finally, the demand for networked security was articulated as a universal remedy through which all of the new threats could be overcome. Networked security thus functioned as an empty signifier, a symbolic representation of different subjects' demands and the common good as such.

Within this larger discourse, military operations were articulated as an integral part of a networked approach, providing support for what was claimed to be mainly a civilian task. Against the background of antimilitarism, the integration of military operations into a networked approach was made possible through a highly ambiguous construction that articulated military operations as simultaneously indispensable and subject to severe limitations. On one hand, German decision makers argued that military operations were a *conditio sine qua non* in (networked) whole-of-government operations, often enabling the application of civilian means in the first place—for instance, in postwar societies. On the other hand, policymakers regularly pointed out (1) the limited utility of the military for the management of new threats like terrorism and (2) that military operations could, for moral reasons, only ever be a means of last resort (*ultima ratio*). This book argues that pre-

---

27. I thank one anonymous reviewer for alerting me to this important limitation; the formulation about the radical potential of these perspectives is the reviewer's.

cisely this highly ambiguous articulation made out-of-area operations possible against the background of sedimented antimilitarist practices. Only because policymakers themselves pointed out the limited utility of military means for the management of the new threats could they credibly claim that those means were nevertheless needed as part of a wider approach. Similarly, policymakers' expression of uneasiness with the use of military means contributed to the impression that participation in military operations was not so much a political decision as a factual necessity to which policymakers only grudgingly conceded, against their own explicit normative convictions.[28]

Precisely in this context, the analytical advantage of a poststructuralist approach vis-à-vis a conventional constructivist one becomes most clearly visible. On face value, decision makers expressing a dislike of military operations seem to demonstrate the continued relevance of antimilitarism. This could be read as evidence supporting the conventional constructivist argument that antimilitarist norms continue to play an important (constraining) role in German foreign and security policy or, in Maull's (2018) terminology, that Germany continues to adhere to a civilian power role. Similarly, Koenig (2020: 91) has recently claimed that "the culture of military restraint" continues "to set important boundaries for the enactment of 'international responsibility.'" In contrast to that claim, a poststructuralist perspective shows how statements expressing a moral aversion to military means are actually employed in favor of, instead of against, military interventions.[29] Thus, a poststructuralist approach reveals how apparently antimilitarist statements serve to undermine military reticence and how the very meaning of antimilitarism is transformed in the process. More broadly, a poststructuralist account can help understand instances of what could be called "paradoxical politics," that is, situations marked by an at least seeming contradiction between rhetoric and policy action.[30]

---

28. The notion of responsibility has received some attention in the study of German foreign and security policy (see Crossley-Frolick 2017; Geis and Pfeifer 2017; Schwab-Trapp 2002; Stahl 2017; Stengel 2010, 2019a).

29. In a similar way, Junk and Daase (2013: 147–48) have pointed out that public acceptance depends on how specific military interventions are framed rather than on an inherent (in)compatibility of interventions with culture per se.

30. The apt term *paradoxical politics* was suggested by an anonymous reviewer.

PLAN OF THE BOOK

This book develops the foregoing arguments in more detail in the following chapters. Chapter 1 draws on discourse theory to develop a theoretical framework for the analysis of discursive change, centered around the notion of hegemony. It sketches an ideal-type hegemonic process, from the disruption (dislocation) of a dominant discourse via discursive struggles between competing projects, to the acceptance, institutionalization, and naturalization of one particular discourse as a new discursive order. In line with discourse theory, the book conceptualizes hegemony as the result of the interplay between (1) the production of a chain of equivalences between previously disparate or even contradictory demands, (2) the construction of an antagonistic frontier between the Self and a radical Other that blocks the Self's identity, and (3) the representation of the chain of equivalent demands by one particularity that, by emptying itself of its particular content, becomes a symbol of a fully constituted society. The chapter pays specific attention to the importance of sedimented practices in endowing certain articulations with credibility. Chapter 2 takes a closer look at discourse theory's ontological and epistemological commitments and explores what these mean for an empirical analysis of processes of hegemonization. In addition, the chapter discusses what explanation means in the context of discourse theory, systematically outlining how such an understanding differs from more conventional, "neopositivist" (P. T. Jackson 2015: 13) notions of explanation. Finally, it explains how the theoretical concepts of discourse theory can be translated into categories for empirical analysis.

Chapters 3–5 provide a detailed analysis of the changing German security discourse since the late 1980s. Chapter 3 examines the old security order that provided the general framework of German security policy during the Cold War. It shows how a positive German identity (as inherently democratic and peaceful) was produced through the double exclusion of (1) Germany's own past and (2) the East, that is, the Warsaw Pact (both of which were articulated as oppressive and aggressive). The discussion of the Cold War order also functions as a foil against which change can be identified. The chapter then turns to the dislocation of the Cold War order at the end of the 1980s and the beginning of the 1990s and the discursive struggles that ensued as a result. It pays particular attention to the rearticulation of the relationship between the discursive elements surrounding peace and military force. The chapter details how arguments for military force to be only a means of last resort (an *ultima ratio*) were reinterpreted in such a way that they actually served to legitimize military operations and how German anti-militarism became transformed in the process.

Chapter 4 analyzes the emergence of comprehensive security as the central concept for the post-unification German policy of conflict prevention. During the 1990s, German decision makers increasingly advocated for the combination of military and civilian instruments to combat armed conflict (what is now known as a networked or whole-of-government approach). The clue about this development is that proponents of comprehensive security picked up demands, originally voiced by members of the Green Party and the peace movement, for more activities in the field of civilian, as opposed to military, conflict prevention and rearticulated them as complementary, instead of an alternative, to military peace operations. This incorporation of competing demands is, I argue, a crucial point that helps explain how military operations became acceptable.

Chapter 5 analyzes the expansion of the discourse of comprehensive security after the terrorist attacks of 11 September 2001 (9/11), from originally the narrower field of conflict prevention to the security discourse as a whole. The main argument developed in this chapter is that as opposed to the US, in which 9/11 proved disruptive, Germany already had the discursive template of comprehensive security ready to make sense of terrorism, as one of the new threats that required a comprehensive or networked approach. As a result, the post-9/11 German security discourse is marked not by upheaval but simply by the expansion of comprehensive/networked security and its establishment as the dominant discursive order (as the general organizing frame for German security policy), thus establishing a new grand strategy. At the same time, using the example of the "war on terror" discourse, the chapter demonstrates how discourses need to be rearticulated to make them credible against the background of the specific sedimented practices of a given society.

CHAPTER 1

Hegemony and Social Transformation

How Discursive Orders Change

This chapter develops a theoretical framework with which to analyze discursive change, drawing on Essex School discourse theory (Laclau 1990b, 1996a, 2005a, 2014g; Laclau and Mouffe 2001). More specifically, the framework (or model, in more conventional terms) tries to provide an explanatory account of processes of hegemonization, that is, the way in which some discourses become widely accepted and institute themselves as a new discursive order (an established, dominant discourse) while others fail (Nabers 2015: 110; Norval 2004: 145). The effectiveness of different discourses is not just of concern to poststructuralists and other researchers primarily interested in discourse. As I will elaborate in more detail below, because poststructuralist discourse theory understands the social as discursive, basically all forms of social change can be analyzed as the result of discursive struggles.

The chapter is organized in two main parts. I begin with a brief illustration of discourse theory's ontological premises,[1] focusing specifically on what a shift toward a discursive conception of the social means for social analysis and foreign policy more specifically. Following that illustration, I develop the aforementioned theoretical framework to explain social change, centered on the notions of dislocation, articulation, and discursive hegemony. The practical consequences that accepting a discursive ontology of the social holds for empirical research are discussed in chapter 2, and chapters 3–5 illustrate the theoretical argument in a detailed case

---

1. For the purposes of this discussion, *ontology* and related terms broadly refer to the metatheory concerned with the nature of being (*Seinslehre*) (see Hofweber 2014; in IR also, Monteiro and Ruby 2009). I return to metatheoretical questions in more detail in chapter 2.

study of the transformation of the German security discourse after the end of the Cold War.

DISCOURSE, THE SOCIAL, AND FOREIGN POLICY

Discourse theory's theoretical starting point and prime concern is to ask how meaning is produced. As opposed to common sense, which usually treats reality as unproblematic and exogenously given, discourse theory directs our attention to meaning-making processes. It claims that how we come to understand the world around us (1) is socially (or, rather, discursively) constructed and (2) has significant consequences for which behavior we commonly deem acceptable in a given situation, including which policy options we consider appropriate, rational, and possible (or not) when dealing with certain policy problems (see Holzscheiter 2014; Milliken 1999).

Like other approaches problematizing forms of naive realism that take reality for granted,[2] poststructuralism is interested in how we come to understand our world and in the political consequences of accepting one represen-

---

2. The claim that "reality" (i.e., how we understand it) cannot be taken for granted—still potentially controversial, at least to readers informed by a neopositivist or critical realist perspective (P. T. Jackson 2010: chs. 3, 4)—is shared by a broad range of theoretical approaches across the metatheoretical spectrum, including not just IR poststructuralism (Campbell 1998; Der Derian and Shapiro 1989; Edkins 1999; Hansen 2006; Walker 1993; Zehfuss 2002) but, equally, feminist (Hooper 2001; Peterson and Runyan 1993; Sjoberg and Tickner 2013; Tickner 1988; Zalewski and Parpart 2008), postcolonial (Barkawi and Laffey 2006; Chowdhry and Nair 2004b; Darby 2009; Doty 1996; Dunn 2003; Said 1979), pragmatist (Friedrichs and Kratochwil 2009; Hellmann 2009a, 2016a), and other "critical" perspectives (for an overview, see Zehfuss 2013), as well as by conventional constructivists (see Hopf 1998; Wendt 1995, 1999) and researchers concerned with perception (Jervis 1976), beliefs (Jervis 2006), framing effects (Bahador et al. 2018; Mintz and Redd 2003), securitization processes (Buzan et al. 1998), problem representation (Sylvan 1998; Sylvan et al. 2005), culture (Weldes et al. 1999), and the exaggeration (Doig and Phythian 2005; Mueller 2005), inflation, and "selling" of threats (Freedman 2004; Holland 2012; Kaufmann 2004; Western 2005), all of whom implicitly or explicitly acknowledge that what we commonly take as a directly accessible, objective reality is at least mediated, if not produced, by various cognitive and social processes. Put simply, not only do different people often "see" the same "facts" differently, but even consensus is the result not necessarily of reality mediating between true and false claims but of shared beliefs, culture, or discourses. For instance, neither the absence of a forceful response to climate change nor the hysteria over terrorism among Western policymakers (see, e.g., Cook et al. 2016; Mueller and Stewart 2018) can be explained by objective problem pressures (i.e., reality as such).

tation of reality over another. The main difference between (most of) the aforementioned approaches and discourse theory is the latter's emphasis on discourse as the central site where meaning, along with the social itself, is produced. Discourse theory assumes that power-laden discursive struggles influence the ways in which we come to understand certain phenomena, the need and urgency to address them, which policy instruments we consider adequate to deal with them, and so on (Nabers 2015).

From this theoretical perspective, discourse is the primary site of processes like securitization and desecuritization, politicization and depoliticization, and the legitimation and delegitimation of certain policies, actors, and so on, as well as the production and reproduction of friends, enemies, and so forth (see Stengel and Nabers 2019). Indeed, much of the "pulling and hauling" (Allison and Halperin 1972: 43) involved in (foreign) policymaking takes place at the level of meaning-making. Understanding exactly what this function of discourse means for social analysis and IR more specifically requires clarifying what the term *discourse* refers to in the context of this book, as that term is often used inconsistently or even unsystematically in IR and political science research (Nabers 2018).

Discourses as Relational Systems of Signification

The central assumption of discourse theory as it is understood here is that all identity is a product of discourse(s). Identity is understood here broadly, as the specific "meaning, sense or signification" ascribed in a certain context to a specific object, practice, symbol, and so on (Torfing 2005b: 153, n1; see also Laclau and Mouffe 1987: 89). Thus, identity here includes but is not limited to the understanding predominant in IR, as denoting a "sense of Self" (Berenskoetter 2010; see also Agius and Keep 2018). German identity, for example, refers to how "being German" or "Germanness" are understood within a specific context. This context is made up of discourses, which can be defined as "relational systems of signification" (Torfing 2005a: 14), that is, specific systems in which the meaning of their parts is constructed through differences—and only through differences (Laclau 2005a: 69; Laclau and Mouffe 2001: 106). As a consequence, from a discourse theoretical perspective, identity is purely differential.

This relational conception of discourse draws on a number of intellectual sources, including the philosophy of Martin Heidegger (see Nabers 2015: ch. 4) and the structural linguistics of Ferdinand de Saussure (2011; see also Howarth 2000). The latter is particularly useful in understanding why difference is fundamental to meaning-making. Important in the context of this study is Saussure's concept of the sign, which is comprised of a signifier (i.e., a sound-image or written word, like "dog") and a signified (an entity or concept to which the sound-image refers—in this case, an animal) (de Saussure 2011: 67). Saussure argues that the relationship between the signifier and the signified is not natural and fixed but "arbitrary" (de Saussure 2011: 67).

Consider the example of an apple. Common sense would suggest that we understand the meaning of the signifier "apple" because the objects it refers to share certain inherent characteristics that make it possible for us to recognize them as part of the same group of phenomena. Similarly, one could argue that most people understand "democracy" to broadly refer to a form of government in which the people (somehow) rule themselves, because that is its nonnegotiable core, its "essence." Although this essentialist view might seem convincing at first glance, it does not actually reflect how language is used in practice (Howarth 1995: 117–18). Not only does any given signifier refer, in practice, to an infinity of signifieds, but the same signified can be referred to by more than one signifier. For example, "apple" can, in principle, refer to an infinite number of different fruits as well as, for instance, wax apples, a computer company, or, in the case of the daughter of Hollywood actress Gwyneth Paltrow, a person so named (see fig. 1).[3] Similarly, an actual apple can be referred to using a number of different signifiers, such as "fruit," "green round thing," "food," and so on.

Because of such variations, Saussure claimed, the meaning a signifier assumes in a given context could not logically be determined by the signified (otherwise it could not change). To the contrary, the specific meaning a signifier assumes in a particular context has to be established in opposition to other signifiers, without any role for the signified in that process. Put simply, we understand what "apple" means in the context of, say, a produce store, precisely because it is not a "pear," a "carrot," or the "shopkeeper." In the different context of, for example, a discourse in information technology, the signifier "apple" assumes a different meaning because it is related to different signifiers: "Apple" refers to technology other than a PC or Android device. Thus, for Saussure, meaning is purely differential, which is what Laclau

---

3. A great number of esteemed journals devote their attention to this matter and similar ones. A reference to the *Daily Mail* (Glennie 2012) shall suffice here.

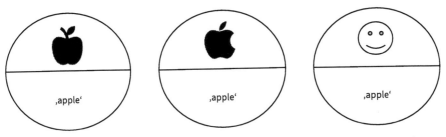

Fig. 1. Signifiers and signifieds. (By Merve Genç, based on de Saussure 2011: 67.)

means when he says that "there are no positive terms in language, only differences" (1989: 68).

This perspective has significant consequences for how we understand political concepts as well. For example, the reason that democracy continues to elude attempts at definition and measurement (see Markoff 2011) is precisely because it has no ahistorical, unchanging essence. What democracy (as an idea) means is the contingent, temporary, and context-dependent result of ongoing discursive struggles over meaning. Otherwise we would be hard-pressed to explain that democracy historically used to be seen as a deviant form of government (Aristotle 1943), is now considered a superior form of government by many in Western societies (infamously, Fukuyama 1989), yet continues to be politically contested (Boyle 2016; Levitsky and Ziblatt 2018; Rensmann et al. 2017). Given this fundamental contingency at the heart of the social, discourse theorists object to the assumption that there is some sort of essence that "drives" society, like the economy in classical Marxism or rationally acting individuals (Stäheli 2006: 254–55).

Importantly, this objection is not only of concern to linguists. If meaning is not determined by the essence of some social phenomenon, practice, or object that is simply there to be discovered, it is the product of discursive struggles in which power plays a crucial role (Howarth 2010: 313; Laclau and Mouffe 1987: 82). Discourses are built in practices of articulation, understood as "any practice establishing a relation among elements such that their identity is modified as a result of the articulatory practice" (Laclau and Mouffe 2001: 105). Articulation links different discursive elements together and thus temporarily fixes their meaning, transforming them into moments within a specific discourse (Laclau and Mouffe 2001: 105–6). Laclau (2005a: 73) has introduced the notion of social demand as the smallest unit from which discourses are built. The term *demand* here refers to both a call for action and a claim (Laclau 2005a: 73; 2005b: 35), including both normative/moral and epistemic (truth) claims.[4] Factual statements are just truth claims that imply the demand to be accepted as true. Since any representation of reality unavoidably has to simplify, it will unavoidably be partial and, as such, political (Bleiker 2001). In Laclau's words, "there is no such thing as a neutral factual description" (2014c: 134; similarly, Horkheimer 1937).[5]

To avoid confusion, it is useful to clarify the relationship between articulations, discourses, and related concepts, before moving on. To begin with, following Laclau and Mouffe (2001: 105), I here use the term *articulation* to refer to both (1) the practice of linking discursive elements and (2) the temporary result of that practice. I use the term *discursive formation* to refer broadly to a connected set of articulations (Nonhoff 2006: 138), while I reserve the term *discourse* for those larger discursive formations that are relatively clearly identifiable, that is, those that can be more or less clearly delineated thematically and/or in terms of their extension in time and space from other discursive formations (Laclau and Mouffe 2001: 105). The term *discourse* refers to both specific discourses and the discursive, that is, the entirety of meaning-making practices (Nonhoff 2006: 32).

I here understand discursive orders as those discourses that have reached a status of near-universal acceptance and become institutionalized to such a high degree that they are taken for granted as quasi-natural. Laclau calls this process sedimentation. As Laclau observes with respect to ideas, sedimented discursive practices are "those crystallized forms that have broken their link with the original intuition from which they proceeded" (2014d: 3).[6] Put simply, although sedimented practices are the contingent result of articulation, they are commonly taken for granted as if they were natural, self-evident, and without alternative. The opposite of sedimentation is what Laclau refers to as reactivation, the process of the denaturalization and contestation of sedimented practices, in which their contingency is made visible again (1990a: 34; 2014d: 3).[7] In the context of this study, what matters most are security orders, that is, the dominant discourses around which security pol-

---

4. I draw here on Nonhoff's (2006: 265) distinction between "stating" (*konstatierend*) and "postulating" (*postulierend*) claims.

5. In IR, this view has been expressed prominently by, for example, Cox (1981) and George (1994).

6. In this process, repetition is an important aspect (Antoniades 2008).

7. A more conventional perspective would call this process "politicization" (see, e.g., Zürn 2014; Zürn et al. 2012).

icy is organized. Security orders construct what is more conventionally called "grand strategy," an actor's (mostly a state's) "theory" of how best to achieve security in a given situation (Posen 1986: 13). They provide an overall definition or reading of the security situation, identify and prioritize the main threats, formulate the best strategy to deal with them, and specify the range of policy instruments to be employed.

Rethinking the Social as Discursive

One important aspect of the Laclauian conception of discourse —and argu-ably the most controversial, certainly to neopositivist social science (P. T. Jackson 2015: 13)[8]—is that it regards the discursive as "co-terminus with the social" (Laclau 2012: 411). This conception combines two aspects. First, Laclau and Mouffe reject the distinction between a (linguistic) discursive and a (non-linguistic) extra-discursive realm, including, for instance, nonlinguistic social practices (Laclau 2005a: 68; Laclau and Mouffe 2001: 107). Laclau and Mouffe insist that all meaningful subjects, objects, and practices are internal to dis-course (Laclau and Bhaskar 1998: 9; Laclau and Mouffe 2001: 107), that what different theoretical perspectives regard as extra-discursive phenomena, such as nonverbal social practices, are, by virtue of being meaningful, inherent to discourse. Laclau and Mouffe explicitly argue that

> any distinction between what are usually called the linguistic and behav-ioural aspects of a social practice, is either an incorrect distinction or ought to find its place as a differentiation within the social production of meaning, which is structured under the form of discursive totalities. (2001: 107)

The Laclauian notion of discourse thus closely corresponds to what Lud-wig Wittgenstein has referred to as a "language game" (1999: 5). In his famous example of the construction of a wall, Wittgenstein illustrates how two builders work together not just on the basis of verbal communication but also through meaningful nonverbal actions, such as the handing of a block or a slab. Both asking for certain construction materials (e.g., a block) and the wordless provision of it are meaningful social action and, as such, part of discourse: "The linguistic and non-linguistic elements are not merely juxta-

---

8. I here refer to those works in political and social science—quantitative and quali-tative alike—that strive to explain the social world in terms of causal relationships, based on a correspondence theory of truth (e.g. King et al. 1994; Mahoney 2010).

posed, but constitute a differential and structured system of positions—that is, a discourse" (Laclau and Mouffe 2001: 108).

This broad conception of discourse distinguishes discourse theory from other discursive approaches, such as critical discourse analysis (e.g., Fairclough 2013; van Dijk 1997, 2006b; Wodak and Meyer 2001), Schmidt's discursive institutionalism (2008, 2010, 2017) or, closer to security studies, "sociological" versions of securitization theory in IR (Balzacq 2015; Bigo and McCluskey 2018; Stritzel 2014), which have a narrower understanding of discourse, as confined to linguistic phenomena. The Laclauian concept of discourse means that we cannot draw a distinction between discourse, on one hand, and, for instance, foreign policy practices, on the other; as a set of meaningful boundary-drawing practices that (re)produce identities, foreign policy has to be seen as inherent to discourse.[9] Such an approach has to reject any arguments that link discourse to foreign policy action in a causal way (e.g., Banta 2013), simply because the two cannot be conceived of as separate.

A second notable aspect of this concept of discourse is that Laclau and Mouffe equally contest the distinction between discourse and the social. To some extent, their argument echoes previous ones about the productive nature of discourse: for instance, Foucault's statement that discourses "form the objects of which they speak" (2002: 54) or Butler's claim that discourses are "performative" (1993: 2). However, Laclau and Mouffe either go beyond (some of) these claims or at least put them in more unambiguous terms. They claim that what we commonly understand to be the social—"the rules, norms, values, cultures, identities, institutions, class structures, gender roles and so on that make society more than the sum total of individual human bodies" (Stengel 2019a: 233)—has no independent existence, no "original presence" (Laclau 2012: 391) that is merely represented in discourse. Instead, discourse is seen as "the primary terrain of the constitution of objectivity as such" (Laclau 2005a: 68), which literally means that the social has no existence whatsoever outside of discourse.

If all meaningful social practices, objects, and so on are inherent in discourse, the distinction between the discursive and the social vanishes. After all, what is the social except the entirety of meaningful subjects, objects, and practices? As opposed to a tree, for instance, which has a physical presence in

9. This broader conception of discourse also informs much of IR poststructuralism (see Campbell 1998; Edkins 1999; Hansen 2006; Hellmann et al. 2016a).

addition to and independent of the meaning we ascribe to it, social phenomena such as class, race, gender, culture, or even political institutions have no independent factual existence beyond their meaning. Thus, while a tree continues to grow even if the entirety of the human race is unaware of its existence, the International Criminal Court ceases to exist if people stop reproducing it in discourse, simply because its whole existence consists of nothing more than "meaning-in-use" (Wiener 2009). Thus, from a discourse theoretical point of view, "ideational factors" (Finnemore and Sikkink 2001: 393) such as ideas (Goldstein and Keohane 1993b), norms (Finnemore and Sikkink 1998; Jepperson et al. 1996), cultures (Berger 1998; Hudson 1997; Lantis 2002b), identities (Banchoff 1999; Checkel 2001; Vucetic 2017), and role conceptions (McCourt 2012; Wehner and Thies 2014) are the contingent, temporary, more or less sedimented (institutionalized) product of ongoing discursive struggles—as are political institutions (Howarth 2000: x, 102; Jacobs 2019;). Thus, despite its name, discourse theory should be understood, above all, as a social and political theory (Nonhoff 2007a: 8).

Let us briefly revisit what this understanding means for the analysis of foreign policy and international politics. First, which phenomena are policy problems (or even threats) to be dealt with and which policy options and instruments are necessary to deal with them is a matter of their context-specific and contingent discursive articulation (Rothe 2015; Stengel 2019b). Depending on how certain subjects, objects, and practices are endowed with meaning, certain policy options will seem more or less appropriate to deal with certain policy problems. Second, as noted above, so-called ideational factors like the values, norms, identities, and role conceptions that influence human behavior are produced, reproduced, and transformed in discourse. Contrary to conventional constructivist scholarship that sees policymakers constrained by largely stable ideational factors (e.g., Berger 1998; Finnemore and Sikkink 1998; Jepperson et al. 1996; Maull 2000; Shannon 2017), a poststructuralist approach will see their precise meaning and their compatibility with specific courses for action as the contingent and changeable result of discursive struggles; that is, the very compatibility between, say, the use of military violence in Afghanistan and German antimilitarism is produced and contested in discourse (Nonhoff and Stengel 2014). As a consequence, what is commonly considered "appropriate" (March and Olsen 1998; Müller 2004b) behavior for a certain state in a specific situation—what is morally acceptable—is a result of discursive struggles.

Since both facts and values are discursively produced, ignoring discourse often means starting with a situation in which a number of potential policy options have already been excluded as unfeasible, improper, irrational, or immoral while others seem like the obvious, self-evident, only courses for action, as IR poststructuralists have highlighted (Diez 2001; Doty 1993; Hansen 2006; Herschinger 2011). Discourses regulate what can be legitimately said and done in a given situation and by whom (Foucault 1971: 8; 1980: 82). As Cohn has famously demonstrated in her research on defense intellectuals, it can sometimes be extremely difficult to act against dominant behavioral expectations in a given situation or even to recognize that alternative courses are available (1987, 1990, 1993). At times, certain options are virtually unthinkable. Negotiating with radical Islamist terrorists, implementing high taxes for corporations and the rich, or bans for sport utility vehicles or guns are examples of potential options that might deal with certain policy problems (terrorism, inequality, climate change, and gun violence, respectively) but that are usually dismissed out of hand, not because of their proven impracticality, but because they appear so extreme or even irrational that they need not even be seriously considered. Discourses do not make it physically impossible to say something, of course, but they certainly delimit what can be said without disqualifying oneself, one way or another, in the eyes of an audience.

Although this aspect of discourse has been most prominently highlighted by Foucault, it is quite easily translated into Laclauian terminology, relying on the notions of credibility and subject positions. Credibility refers to a statement's compatibility with established (sedimented) discursive practices (Laclau 1990a: 66). Above all, the credibility of an articulation depends on whether it clashes with the "ensemble of sedimented practices constituting the normative framework of a certain society" (Laclau 2000: 82), the "basic principles informing the organization of a group" (Laclau 1990a: 66). Any statement clashing with these sedimented practices will likely be rejected.

Importantly, this rejection includes both normative/moral and factual/epistemic statements. Articulations can fail to gain acceptance because they clash with moral (including legal) principles. In the context of German foreign policy, the most relevant set of normative sedimented practices are arguably the traditional "normative foundations" (Bosold and Achrainer 2011) of German foreign policy: most notably, antimilitarism, multilateralism, and Western integration (e.g., Baumann 2011; Maull 2014; Risse 2007).[10]

---

10. I am not claiming that antimilitarism, for example, has a fixed essence.

As I discuss in more detail in chapter 3, these traditions are bound up with myths (Bevir and Daddow 2015: 279)—in the German case, primarily the myth of a "new," post-1945 Germany that had nothing in common with its authoritarian predecessors. Policy demands that openly clash with foreign policy traditions will likely be met with rejection. Also, demands incompatible with widely accepted truth claims ("facts") and the relations between them—what political psychologists would call "causal beliefs" (Tetlock 1998: 641)—will likely fail. For instance, US president Donald Trump's reported 2017 proposal to nuke hurricanes was met with ridicule because it is nonsensical from a practical point of view.[11]

On the flip side, statements that resonate with and explicitly draw on sedimented discursive practices will have an immediate, common-sense appeal (Nonhoff 2001). In that respect, discourse theory converges with the understanding of political psychologists who stress that propositions are most likely to be accepted when they "fit . . . more general beliefs" (Jervis 2006: 651). For instance, that arguments related to national security often prove to be a particularly "powerful discursive weapon" (Rothe 2015: 46) is the result of sedimented discursive practices according to which national security has to take precedence over alternative concerns. Because hegemonic projects have to incorporate sedimented discursive practices to gain credibility, discursive orders are not built from scratch. Rather, they combine old and new elements, weaving them together in a form of "discursive 'bricolage'" (Stengel 2019b: 304) and transforming their meaning as a result of their changed differential combination.

Similarly, intertextual references articulating that demands voiced in other, highly sedimented discourses or individual texts are in line with one's own demands can strengthen credibility (Hansen 2006: ch. 4). Widespread and highly sedimented "grand discourses" like the ones on justice, freedom, welfare, and so on articulate widely accepted basic values. References to such values as democratic participation, human rights, the rule of law, a free market economy, or the self-determination of peoples can increase the credibility of a particular project by tying it into established liberal discourses (on liberalism, see Freeden 1996, 2003).

In the same vein, demands for military interventions draw on sedi-

---

Exactly what these signifiers mean is always a matter of their contingent contextual articulation.

11. https://www.axios.com/trump-nuclear-bombs-hurricanes-97231f38-2394-4120-a3fa-8c9cf0e3f51c.html

mented gendered and racialized discursive patterns for credibility. Thus, arguments about the need to help women and children (see Carver 2006), prominent in interventionist discourses (Darby 2009; Koddenbrock 2012; Masters 2009; Sabaratnam 2018; Shepherd 2006), are intuitively convincing because they resonate with established notions of masculinity and femininity that, in turn, are closely associated with certain characteristics, like strong/weak, active/passive, protector/victim, rational/irrational, and sober/ emotional (with the prior, masculine option being privileged).[12] Numerous studies have highlighted the intimate links between gender constructions and the legitimation of armed forces and military violence.[13] One prominent example is the so-called logic of masculinist protection, by which decision makers invoke the figure of the soldier as a selfless hero to solicit obedience from the public in exchange for protection, thus legitimizing not only the armed forces but security political decisions, while delegitimizing criticism, which appears ungrateful (Young 2003). Similarly, arguments for interventions usually draw on what Baaz and Stern (2009: 496) have called the "universalized storyline of gender and warring," according to which men and masculinity are intimately bound up with war, while women and femininity are closely linked to peace (see in particular Elshtain 1982).

Bound up with gendered discourses, demands for the (often predominantly white) "modern" Western Self to help (and simultaneously patronize: see Campbell et al. 2011; Richmond 2006; Spivak 1988) a (predominantly nonwhite) non-Western, "traditional," somehow backward and passive Other often seem credible to Western audiences because they map onto the "older binary between colonizer and colonized" (Chandra 2013: 485).[14] Overall, the appeal (in the West) of not just military interventions but also often unjust and/or patronizing development or economic policies seems

---

12. This point has been discussed at length in feminist IR scholarship: see, e.g., Carver 2008; Enloe 2016; Hooper 2001; Peterson 2010; Peterson and Runyan 1993; Shepherd 2006, 2007; Sjoberg 2010; Tickner 1988, 1992; Weber 1999, 2014a, 2014b; Wilcox 2014. The general theoretical underpinnings have been formulated by Butler (1990, 1993) and Connell (2005).

13. The literature on the topic is quite substantial (see Åhäll 2012; Belkin 2012; Cohn 1987, 1990, 1993; Eichler 2014; Enloe 2000; 2016; Goldstein 2001; Hooper 2001; Hutchings 2008; Sjoberg 2011; Stachowitsch 2013; Wibben 2018).

14. This topic has been discussed at length in postcolonial scholarship in IR: see, e.g., Baaz and Verweijen 2018; Barkawi and Stanski 2012; Blaney and Inayatullah 2018; Chowdhry and Nair 2004b; Crawford 2002; Darby 2009; Doty 1996; Duffield 2005; Dunn 2003; Inayatullah 2014; Jabri 2013; Muppidi 2012; Sabaratnam 2018; Vucetic and Persaud 2018.

difficult to explain without taking into account how arguments in favor of them draw on established discursive patterns around gender and modernity/coloniality that depict the non-Western Other as "powerless, ignorant or delinquent" (Hutchings 2019: 5) and that reproduce a hierarchy between a "modern" West and a "traditional" or otherwise backward non-West.[15] Germany is no exception here. Although German colonialism has often been and continues to be dismissed as a rather irrelevant period (mainly due to its comparatively short duration), recent historical scholarship has demonstrated that German colonialism had a much more significant effect not only on colonized peoples but on Germany itself (e.g., Conrad 2012; Friedrichsmeyer et al. 1998; Perraudin and Zimmerer 2010). Eurocentrism and colonial arguments continue to be reproduced not only in German media discourse but even in educational practices in school (Macgilchrist et al. 2017), which is precisely how arguments drawing on colonial rationality (e.g., for interventions) appear convincing, because they are familiar in light of sedimented practices.

As a result of the specific discursive construction of certain issues, some policy options are immediately appealing, while others are excluded from the start. For example, in the German security discourse, the September 2001 terrorist attacks were articulated as motivated not by rational means-ends considerations but by an irrational ideology (Entman 2003: 424; for a discussion of similar arguments in the US, see R. Jackson 2007: 409). This articulation is reflected, for example, by Chancellor Gerhard Schröder's claim that the attacks were directed against the "civilized world" and its values rather than a response to specific policies (14/187, 19 September 2001: 18301). In articulating 9/11 as an uncalled-for attack from the outside on a principally benevolent ("civilized") Self, Schröder externalizes the problem. As a consequence, fighting terrorism appeared as the only viable response, while alternative policy options, such as negotiations or even the consideration of some arguments of

---

15. The modernity/coloniality dichotomy has received significant attention in postcolonial theory. I here understand coloniality, with Tlostanova, as "a racial, economic, social, existential, gender and epistemic bondage created around the 16th century, firmly linking imperialism and capitalism, and maintained (though reconfigured) since then within the modern/colonial world" (2012: 132). The important point here, in my opinion, is the emphasis on elements of continuity between colonial times and the present, on ways in which the colonial past continues to exert influence in the present, through, for example, the reproduction of hierarchies between former colonizers and former colonized, relations of exploitation, epistemic violence (Brunner 2017), and so on. For a more general theoretical discussion see, *inter alia*, Quijano 2007; Mignolo 2007; Mignolo and Escobar 2010.

al-Qaeda leader Osama bin Laden, were dismissed as irrational. Foreign Minister Joschka Fischer claimed in September 2002,

> We will not be able to negotiate with Osama bin Laden. What would [even] be the subject of negotiation [*Worüber wäre da zu verhandeln*]: that they should kill fewer innocent people, that they should refrain from destroying Israel, that they should desist from terrorism? None of that will work. That is why the bitter consequence is [this]: We will overpower this terrorism. We will have to defeat it. (14/253, 13 September 2002: 25593)

As a result of the construction of the terrorist Other as irrational and devoid of any rational reasons for the attack, fighting and defeating terrorism appears to be the only viable option. More often than not, that one particular representation of reality becomes accepted as "the truth" or that one course for action becomes regarded as the "obvious" or only choice does not mean that they objectively are. For instance, the above representation of reality articulated by Fischer is but one way of understanding that situation. Contrary to the articulation of the terrorist Other as irrational and beyond reasoning, bin Laden, in his "Letter to the American People," claimed that the reason behind the attacks was "because you [the US] attacked us and continue to attack us" (bin Laden 2002), citing, among other concerns, US support for Israel, Somalia, for Russia in Chechnya, for India in Kashmir, and for authoritarian governments in Islamic countries.

My argument here is not that 9/11 was morally justified or that giving in to al-Qaeda's demands—among them the call to conversion to Islam (bin Laden 2002)—would have been a better option. Rather, my point is that dismissing any and all of bin Laden's arguments out of hand, as irrational and beyond reasoning, has unintended side effects. Most notably, claiming that 9/11 is rooted in al-Qaeda's extremist ideology alone makes it nearly impossible to even entertain the idea that past (and present) Western policies might play a role in the creation of anti-Western sentiment. This claim stands in contrast to a number of studies that have highlighted the role of the West in producing the very problems that are commonly said to originate outside it, including, most notably, the rise of al-Qaeda and the Taliban (Khalilzad and Byman 2000; Mamdani 2002; Muppidi 2012; Patman 2015). Even if one does not share Johnson's argument that 9/11 should be seen as "blowback" for past US policies (Johnson 2004a, 2004b), it stands to reason that, for example, continued Western support for authoritarian regimes like

Saudi Arabia, ill-advised military interventions, and the sustenance for an unfair world economic order are not likely to reduce anti-Western sentiment (Chandler 2010; Hoffmann 2001; Inayatullah 2013a: 334; Mamdani 2005). However, such arguments fall on deaf ears if hatred against the West is explained away as being exclusively (instead of partly) the result of an extremist ideology. If one presumes to already know the "real" reason behind the attacks (hatred for "our" values), there is a strong incentive to dismiss any reasons given, as mere rhetorical window dressing or the ramblings of a lunatic. Moreover, by taking a madman's arguments seriously, one edges uncomfortably close to madness. Yet the construction of the terrorist Other stands in the way of seriously questioning the role of Western policies in the reproduction of violence.

The credibility of a given statement is influenced not only by its content but by the subject position of the speaker, which specifies the latter's position within the discursive order. Discourse is far from a level playing field, and depending on their subject positions, subjects will be able to speak with authority on certain matters and within a specific context (say, as a doctor in a medical discourse) or not (Foucault 1971: 17, 2002; Laclau and Mouffe 1987: 82). This authority is regulated, above all, by institutions (including public offices), understood here as sedimented practices that offer subject positions endowed with certain rights and privileges (Laclau 1996d: 43).[16] For instance, in matters of security policy, high-ranking Bundeswehr officers are commonly considered experts and, as such, authorized to speak on these matters. At the same time, even the inspector general of the Bundeswehr, the highest-ranking officer in the German armed forces, is not authorized to speak in the German Bundestag, because that privilege is reserved for members of parliament (MPs), the federal government, and the upper house, the Bundesrat (art. 43 GG).

Aside from formal authority that derives from holding an official post, the credibility of individual speakers can vary because they occupy more informal subject positions due to, for example, their personal standing within a certain group (or lack thereof). One example is the role in the Bundestag of the left-wing party Die Linke, whose members are usually not taken seriously in matters of security policy by the more established parties, due to

---

16. As opposed to continuing claims that discourse theory "failed fully to take on board the institutional dimension of politics, resulting in a view of politics that is largely devoid of institutions" (Panizza and Miorelli 2013: 302), discourse theory allows for an incorporation of institutions into its framework (Jacobs 2019).

a mixture of what is said to be the party's populist and utopian attitude as well as its history. Die Linke is the successor of the Socialist Unity Party (Sozialistische Einheitspartei Deutschlands, SED), which ruled the German Democratic Republic (GDR) until 1989/90 (Hough 2000: 127), and Die Linke's critics argue that the party has failed to reappraise and deal with its own authoritarian past, which disqualifies its members from legitimately speaking up on topics of oppression and violence, in particular. As a consequence, statements by Die Linke MPs are often dismissed out of hand, without any engagement with the contents of the argument (see, e.g., zu Guttenberg, 16/49, 19 September 2006: 4816).

By providing a particular and always partial interpretation of reality, discourses make some courses for action more likely than others. Discourses regulate not only what can be said and done but also who is authorized to say and do certain things in a given context. The discourse theoretical conception of the subject is bound up with the possibility of social change.

DISLOCATION, ARTICULATION, AND HEGEMONY: HOW DISCURSIVE CHANGE WORKS

In principle, we can analytically distinguish between two types of discursive change, depending on their extent. Microlevel discursive change is limited to change within an existing discursive order, whereas macrolevel discursive change refers to transformation in the dominant discursive order as a whole (Stengel 2019b).[17] In this context, the distinction between microlevel and macrolevel change is heuristic; in reality, discursive change is best understood as a continuum, not a binary category. Moreover, the empirical distinction between microlevel and macrolevel change has to be seen as a (performative) interpretation rather than the objective description of an independent reality (see chapter 2). Nevertheless, the distinction is introduced here to be able to distinguish the transformation of whole discursive orders from change within an existing order. This distinction is relevant

---

17. The use of *micro* and *macro* in this context should not be confused with the use of these terms in sociology, where microsociology is the "detailed analysis of what people do, say, and think," while macrosociology is concerned with far-reaching and long-term social processes (Collins 1981: 984). Microlevel discursive change can affect whole societies, because the reference point is not the size of the group but the extent to which a discursive order is transformed.

because attempts to rearticulate discourses will face more or less obstacles depending on the extent of change. As a rule of thumb, far-reaching discursive changes rearticulating entire discursive orders will be more difficult to achieve, because they go against and seek to transform conventional wisdom and/or widely accepted normative commitments, whereas rearticulations at the microlevel primarily have to be credible against the background of sedimented discursive practices.

Subjectivity, Dislocation, and Rearticulation

In contrast to more rationalist or conventional constructivist approaches in political science and IR (e.g., Elster 1988; Scharpf 1997: 19) that take some "pre-social individuality" (Wendt 1999: 179) as a given, poststructuralists usually stress processes of "subjectivation" (Laclau 2000: 78), that is, processes through which individual human beings are "made" into subjects by becoming part of society (Foucault 1982: 777; see Howarth 2013; Nabers 2018).[18] Concentrating on subjectivation means to question how a given subject's identity—its sense of Self—is constituted in discourse. Understood this way, identity refers to the ways in which an individual or collective subject is signified (differently) in different discursive contexts.

Discourses provide subject positions that human beings can embrace or with which they can identify. Individual human beings can identify with a number of different subject positions, depending on the context. For example, the same human being can be a soldier, mother, political activist, customer, or voter, to name just a few. All of these different subject positions are associated, within a specific discursive context, with certain characteristics, behavioral expectations, rights, responsibilities, and so on. The most obvious example here is gender discourse, which intersects with thematic discourses and influences standards of appropriateness along gendered lines. For example, a soldier (coded masculine) is commonly expected to be rational, decisive, strong, tough, brave, emotionally detached, and in control, while a mother (coded feminine) is commonly expected to be caring and

---

18. This process is sometimes referred to as "subjectivization" (Rancière 1992), "subjectification" (Mansfield 2000: 62), or "objectification" (Foucault 1982: 777). The distinction between human beings and subjects is purely theoretical, as "mere," presocial human beings do not occur in practice; individuals are always already embedded in discursive structures and, as such, always already specific subjects (Nonhoff and Gronau 2012: 123).

nurturing toward her children (Basham 2016; Higate 2003; Hooper 2001; Miller 2005).[19] Similarly, political leaders are commonly expected to be rational, decisive, and strong, which is why being seen as "soft" (e.g., on terrorism) can undermine a politician's authority (Jones 2004; Millar 2018; Tickner 1988; Young 2003). Gender is but one example of how subjectivation works, but it illustrates how powerful discourses can be, not through physical coercion, but by creating expectations for what is "proper" conduct for a certain class of people. Again particularly instructive in this context is Cohn's research (1987, 1990, 1993) on how constructions of masculinity and femininity regulate what can (not) be legitimately expressed in a national security setting.

This rather far-reaching influence of discourse brings up the question of what is commonly referred to in IR as "agency" (e.g., Doty 1997; Wendt 1987; Wight 1999b), meaning the possibility of deliberate, purposeful action by human beings, undetermined by social structures. Needless to say, the idea that subjects are discursively produced has not been met with universal acclaim, primarily because it clashes with more conventional political science explanations in which policy outcomes are usually analyzed by focusing on intentional actions (within the constraints provided by institutions and/or other structural features, like anarchy) of rational actors endowed with free will (e.g., Elster 1988; Keohane 1984; Moravcsik 1997; Scharpf 1997: 19; Waltz 1979).

Contrary to critics' claims that poststructuralism denies the possibility for strategic action and reduces human agents to "passive bearers" of discourses (Stritzel 2012: 551; similarly, Bevir and Rhodes 2003: 43), which would ultimately deny the possibility of social change (Laclau 1990a: 35), discourse theory does not dispense with human agency. There are at least two reasons for why human beings are not mere implementation automats for discursive structures. The first point concerns the multiplicity of individual identity. Any human being is always embedded in and constituted as a subject by multiple discourses. A single person can identify with a number of different subject positions, and the behavioral expectations associated with each different subject position can sometimes interfere with and contradict each other (Eberle 2019: 19).

The second aspect is commonly discussed under the heading of "unfix-

---

19. I am simplifying here for illustrative purposes. Military masculinity/ies is a good deal more complicated and contradictory than the example suggests (Belkin 2012; Henry 2017).

ity" or "dislocation." It is the basis for the possibility of social change. Contrasting Saussure's structural linguistics with poststructuralism is illustrative here. Although Saussure stresses the arbitrariness of the sign, he presumes a strict isomorphism between a differential system of signifiers and a corresponding differential system of signifieds (Laclau 1989: 68), in which each single signifier in a discursive formation corresponds to one single signified (Torfing 1999: 87–88). Such a system leaves no room for change, because it is eternally fixed. In contrast, poststructuralist discourse theory emphasizes the simultaneous subversion and exceeding of any particular identity (Laclau and Mouffe 2001: 104). On one hand, every signifier can, in principle, be attached to every signified (see the example of the signifier "apple" discussed earlier in this chapter). On the other hand, any signified can be equally represented by an infinite number of signifiers.

In Laclau and Mouffe's (2001: 104) words, each identity is "overdetermined." When the meaning of a specific signifier is fixed within a specific discursive context, it assumes one particular meaning out of a principally infinite number of possible meanings, while all other meanings are excluded. The signifier (a discursive element) is "reduced to a moment" of a particular discourse (Laclau and Mouffe 2001: 106); its meaning is narrowed down and (temporarily) fixed. As a consequence, any specific connection between a signifier and a signified is context-dependent and rests on the exclusion of a plethora of potential other meanings. This also applies to discourses as totalities, which are made up of moments. Thus, from the perspective of any specific discourse, there is always a "surplus of meaning" (Laclau and Mouffe 2001: 111)—all other potential meanings that are excluded in a specific context. The entirety of excluded potential meanings is what Laclau and Mouffe call the "field of discursivity" (2001: 111).

In practice, then, meaning can always only be temporarily fixed. The signifier "freedom," for example, assumes a very specific meaning in the context of neoliberalism and an altogether different one in a Marxist discourse (Biebricher and Johnson 2012: 205; Wolff 2011). Any claim to what freedom really means remains principally open to being questioned and challenged. Both Marxism and neoliberalism ultimately fail to monopolize the meaning of freedom and thus remain vulnerable to rearticulation (Laclau and Mouffe 2001: 106-7). Neither the meaning of a single signifier nor that of a discursive totality can ever be completely fixed (Laclau 1989: 69). As a consequence of this inability to monopolize meaning, of the "impossibility of closure" (Laclau and Mouffe 2001: 122), any and all discursive structures are characterized by a "lack" (Laclau 1994: 3). Any structure is marked "an ineradicable distance from itself" (Laclau 1990a: 60).

In *New Reflections on the Revolution of Our Time*, Laclau introduced the term *dislocation* to refer to this chronic, "constitutive" disruption of all discursive structures (1990a: 39; see also Laclau 1997: 302).[20] Because discourse theory conceives of the social as coextensive with discourse, the social as such—cultures, identities, institutions, and so on—is equally marked by a chronic incompleteness. Thus, a poststructuralist account would highlight the context-dependent, contingent, contested, inconsistent, and sometimes even contradictory nature of the ideational factors that conventional constructivists usually take to be rather stable (Nonhoff and Stengel 2014).[21] Social identities in general are marked by the unstable, dislocated nature of all discursive formations, since "there is no social identity fully protected from a discursive exterior that deforms it and prevents it becoming fully sutured" (Laclau and Mouffe 2001: 111).

Dislocation is also mainly why the charge that poststructuralism leaves no room for agency is misguided. Because no discourse can fully fix meaning, any discourse will unavoidably encounter events that it cannot incorporate, domesticate, or explain, which further contributes to dislocation. In this moment of dislocation, when the discursive structure is disrupted by events it cannot incorporate, the links between different moments of the discursive formation are loosened. As a result, these moments transform into "floating signifier[s]," open to rearticulation (Laclau 1989: 80).[22] The subject comes into play here. Because the structure is always dislocated to some extent, it fails to fully determine the subject (Panizza and Miorelli 2013: 2). The subject always has some degree of "wiggle room." In this way, the dislocation of the structure is "the source of freedom" (Laclau 1990a: 60).

---

20. For highly insightful discussions of dislocation, see Nabers 2017, 2019.

21. Some constructivists too have pointed to the contested nature of ideational factors like identity or culture (e.g., Biehl et al. 2013: 12; Katzenstein 2003: 738). For critique of constructivists' static conception of ideational factors, see Baumann 2006; Bially Mattern 2005; Flockhart 2016.

22. I am not claiming that discourses are momentarily disrupted by external shocks or crises. What I mean by a dislocatory event is the "pure event" that, precisely because it cannot be represented, only appears as an "interruption in the normal course of things, a radical dislocation" (Laclau 1996b: 73–74, italics removed; see also Lundborg 2012). Once a pure event has been represented—e.g., as a shock or a crisis (on the construction of crises, see Nabers 2015, 2017; Widmaier et al. 2007)—and thus becomes a discursive event, it is simply another moment in the discursive structure, with a specific meaning (Marchart 2007: 74).

This freedom is a mixed blessing at best. Since social identities are produced in discourse, the lack at the heart of any discursive structure also means that the subject, while having some degree of freedom, has to cope with a disrupted identity. Like the structure as a whole, the subject's identity (collective or individual) is a *"failed* structural identity" (Laclau 1990a: 44, italics in original). As a consequence, the subject experiences its freedom as distressing. Incorporating insights from Lacanian psychoanalysis,[23] Laclau conceived of the subject as "split" (1995: 159) or "traumatized" (Panizza and Miorelli 2013: 315; Torfing 2005a: 17)—a subject *"condemned* to be free" (Laclau 1990a: 44, italics in original).

The subject deals with this lack at the heart of all identity by embracing different attempts to "re-suture" the dislocated discursive structure (Norval 2004: 142) or, more precisely, with subject positions offered by specific discursive formations.[24] Laclau uses the term *identification* for the subject's decision to embrace a specific subject position provided by a discourse (2000: 58). Because all discursive structures are dislocated to some extent, the subject can always only gain an incomplete identity as a result of identification (Laclau 1990a: 44). Rather than being able to attain a stable identity, the subject is constantly "literally 'forced' to act and identify anew" (Howarth 2005: 323), condemned to forever chase a full identity that remains out of grasp. Although complete fixation, "closure," remains an unattainable goal, the subject cannot escape from ongoing attempts to reach it (Laclau and Mouffe 2001: 112, 22).[25]

Two aspects are important to note in this context. First, the subject's decision takes place "in an undecidable terrain" (Norval 2004: 143). Here, undecidability means not that a decision is impossible but, rather, that the decision is not determined by the structure. In Laclau's words, it refers to "that condition from which no course of action necessarily follows" (1996b: 78).

---

23. For IR studies informed by Lacanian psychoanalysis, see, in particular, Eberle 2019; Epstein 2011; Solomon 2014; Zevnik 2016.

24. Subjects cannot always choose to (or refuse to) identify with subject positions; they might not be authorized to speak in a particular discourse or may be physically absent. Still, their particular articulation can legitimize certain policies. For instance, the construction of helpless women and children can help justify armed interventions (Shepherd 2006: 20).

25. This view is quite similar to arguments about ontological security, according to which subjects strive for a stable, continuous, whole identity over time (Eberle and Handl 2020B; Mitzen 2006; Steele 2008). However, a discourse theoretical perspective places emphasis on the inescapable "basic ambiguity at the heart of all identity" (Laclau 2000: 58).

Decisions in an undecidable terrain are radical insofar as they "cannot appeal to anything in the social order that would operate as its ground (otherwise it would not be radical)" (Laclau 1994: 4). From a Derridean perspective, this situation would be called "aporia," a state in which no single, obvious way forward exists on the basis of normative principles, because neither of the available options would "satisfy all rules, principles, and obligations at the same time" (Zehfuss 2018: 46). In such situations, the subject is forced to identify with one of a number of competing interpretations of reality, without being guided by, for example, normative principles (because that would mean that it is not really a decision) that would unambiguously identify a single "right way forward" (Zehfuss 2007: 262). For Laclau, then, the moment of decision is the location of politics and the political, understood as the "practice of creation, reproduction and transformation of social relations" (Laclau and Mouffe 2001: 153).[26] Politics, as Critchley aptly puts it, is "the realm of the decision" (2004: 113; see also, recently, Leek and Morozov 2018).

This moment of dislocation is precisely where agency, in the traditional IR sense, is located. Howarth observes,

> If both the structure and subject are marked by a fundamental foreclosure—an impossibility which becomes evident in moments of disruption—then in certain conditions the subject is able to act in a strong sense: to identify with new objects and ideologies. This moment of identification is the moment of the radical subject, which discloses the subject as an agent in its world. (2010: 314)

The failure of the discursive structure, the "distance between the undecidable structure and the decision" (Laclau 1990a: 30), makes intentional action possible and, at the same time, compels the subject to act.

Notable, second, is that the incompleteness of the structure does not equal its complete absence. The subject is not determined by the structure, but it is not completely unconstrained either. The subject's decision takes place within a setting of sedimented discursive practices, not all of which will be entirely dislocated (Laclau 1990a: 35; Nonhoff and Gronau 2012). How much wiggle room the subject has in a given situation—and therefore

---

26. I understand "politics" here as concerned with specific decisions in concrete contexts (at the ontic level), while I understand "the political" as an ontological horizon for all possible decisions at the ontic level (Laclau 2012; Laclau and Zac 1994; Nabers 2015).

the possibility for change—depends on the extent to which the discursive structure is dislocated. As a rule of thumb, "the more dislocated a structure is, the more the field of decisions not determined by it will expand" (Laclau 1990a: 39–40, 66), that is, the more space opens up for political attempts to rearticulate the structure, including deliberate, strategic action.

The chronic dislocation of all discursive structures makes social change possible. The openness of all social structures means that they are vulnerable to rearticulation, and the ambiguity and instability of meaning means that specific representations are always open to be challenged. That is particularly the case with floating signifiers such as "freedom," which have a different meaning in different discourses and, as a result, remain more contested than other signifiers (Laclau 1997: 306). As a consequence of dislocation, discursive structures are always in flux to some extent, subject to ongoing processes of rearticulation, in which the discursive elements are rearranged. This rearticulation is how change at the microlevel functions: because subjects are constantly involved in articulatory practices (through speech and nonverbal action), discursive elements are constantly rearranged in these ongoing practices.

Whether specific attempts to rearticulate practices at the microlevel are accepted or provoke resistance depends, above all, on their credibility, that is, their compatibility with sedimented practices. Since sedimented practices do not always form a coherent whole, attempts at rearticulation can be successful by drawing on certain sedimented practices while challenging others. One example for such change, in which moments within a discursive order are rearticulated without the establishment of a new order, is the expansion of the category of new threats. As the empirical chapters of this book trace in much more detail, at the heart of the post-unification security discourse was the claim that the old Soviet threat had been replaced by new threats like armed conflicts, terrorism, and state failure. The establishment of the category of new threats itself was the result of macrolevel change, but even after its establishment, that category remained subject to rearticulation, as discourse participants continued to articulate additional phenomena as part of the new threats. Over time, new phenomena like piracy or cyberterrorism were incorporated into the discourse and integrated into the category of new threats. This expansion of the category was possible because discourse participants could make credible arguments that new or previously neglected phenomena were actually causally linked to previously known new threats, with, for instance, cyberterrorism and cybercrime being digital versions of terrorism and organized crime or with piracy contributing to and feeding off state failure and poverty. The success of such relatively limited rearticulations, in which new elements are added to an existing discursive order without the latter being overturned, depends, above all, on their credibility vis-à-vis the set of sedimented practices, that is, on their compatibility with the existing order.

However, processes of identification do not necessarily have to lead to change. In principle, we can distinguish between a political response to dislocation, in which established social relations are publicly contested and potentially rearticulated (Laclau 1990a: 35), and an ideological one, which aims to "repair or cover over the dislocatory event before it becomes the source of a new political construction" (Glynos and Howarth 2007: 117). In line with this distinction, political demands challenge established practices (Glynos and Howarth 2007: 115). In contrast, ideology refers to the moment of closure in which a specific project temporarily and partially fixes meaning (Laclau 2006: 103, 14). Ideology conceals the contingency of the social, because it necessitates that a contingent and unavoidably incomplete identity presents itself as fully constituted, self-transparent, and quasi-natural (Laclau 2014b: 15–16). Ideological in that sense, for instance, is speaking about "the West" as if it was unproblematic and had an existence independent of articulatory practices reproducing it (see Hall 1992; Hellmann and Herborth 2016; P. T. Jackson 2006; Klein 1990). Laclau uses the terms *myth* and *social imaginary* to refer to ideological representations, with myth simply denoting "a principle of reading of a given situation" (1990a: 61). For Laclau, *ideology* is not a pejorative term (2006: 114); in fact, given that representations will unavoidably be partial, ideology is inescapable, because the only alternative to ideological representations would be to have "no meaning at all" (2014b: 16).

In sum, discourse theory conceives of all social structures and identities as partially dislocated, which makes them vulnerable to rearticulation and forces the subject to identify with attempts to repair the discursive structure, promising (the illusion of) a full identity. Because structures are dislocated, the subject is forced to fill in the gaps through constant acts of identification, thus reproducing and gradually transforming the discursive order. The success of minor rearticulations depends, above all, on their credibility, that is, the extent to which they resonate with established sedimented practices. Things are more complicated when it comes to macrolevel discursive change, because it seeks to change a significant part of the sedimented discursive order itself. While credibility also plays an important role here, attempts to replace or transform a discursive order—so-called hegemonic projects—need to have a much broader appeal.

Hegemony and the Transformation of Discursive Orders

I have specified, in a general fashion, how agency and social change can be conceptualized from a discourse theoretical perspective. How whole discursive orders are transformed remains for analysis. I presume here that the more far-reaching attempts to rearticulate established discourses are, the more obstacles they will face. Humans are creatures of habit, and constantly questioning social practices would simply paralyze people. If people were to constantly, say, question the necessity of stopping at a red light, calculate the risk of getting caught riding the bus without a ticket vis-à-vis the costs of buying one before every ride, or think through the potential consequences (for the environment, workers' rights, one's own health, the continued existence of patriarchy, racism, capitalism, etc.) of buying a certain product, life would grind to a halt. As a consequence, much of social life is organized around taken-for-granted, routinized social practices, which we often simply reproduce without question (Glynos and Howarth 2007: 104).

How, then, does discursive change at the macrolevel happen? As opposed to, for instance, either a critical realist approach arguing for a "'rump' materialism" that constrains which articulations (constructions) of material reality are possible (Banta 2013; Wendt 1999: 110) or some poststructuralist approaches that distinguish a discursive from an extra-discursive realm (e.g., Diez 2014), discourse theory offers an explanation inherent to discourse. Central here is the notion of hegemony, the process by which some hegemonic projects—attempts to establish the dominance of a specific discursive formation built around a broad, comprehensive demand (Nonhoff 2019: 76)—manage to prevail over competing projects. In its broadest sense, the term *hegemony* refers to the "operation of taking up, by a particularity, of an incommensurable universal signification" (Laclau 2005a: 70). Put simply, it means that a specific demand (a particularity) assumes the function of a symbol not only of a broad range of social demands but also of the "absent fullness of society" (Laclau 2005a: 71),[27] that is, the (unattainable) ideal of a

---

27. "Absent fullness" means that although full closure is impossible (hence "absent"), the ideal of a fully constituted society is still present as a desire, as an unattainable goal that nevertheless remains out of reach.

perfect (i.e., fully constituted) society (the universal) in which all demands are fulfilled. At the same time, if hegemony is successful, one particular way of understanding the world (one particular discourse) establishes itself as the only valid understanding of the world (a universal).[28]

Ideal-typically,[29] a hegemonic process involves a number of different steps, beginning with the dislocation of an established discursive order and ending with the institutionalization of a new one. Figure 2 provides a simplified illustration of this process, which is a continuous cycle rather than a closed system with a fixed beginning and end. This process is not as straightforward in reality as the figure suggests, nor are the different steps of the process as neatly separable or as uncontested; they are much more characterized by discursive struggles and contradictions. The figure primarily serves as a heuristic device to make the process more easily understandable.

Hegemony begins with the dislocation of an established discursive order. As a result of dislocation, its legitimacy is increasingly questioned by subjects, and discursive struggles emerge between the advocates of different hegemonic projects to replace and/or fix the disrupted order. Subjects will identify with one or another hegemonic project in order to repair their own disrupted identities. At some point, one particular project will gain widespread acceptance, replacing the old discourse. Along with this replacement goes the erasure of the project's origin in political struggles, as the project becomes naturalized as the single valid (instead of just one possible) way of understanding the world.

A number of "design characteristics" influence a hegemonic project's chance of success. To begin with, the project has to demonstrate what Laclau terms "its radical discontinuity with the dislocations of the dominant structural forms" (1990a: 67). In simple terms, a new hegemonic project has to "learn from the failure of previous discourses" (Nabers 2009: 197). After all, hegemonic struggles emerge because the dominant order failed to incorporate and make sense of certain dislocatory events, and any project aspiring to become the new dominant order has to make a convincing case that it would fair significantly better. Beyond offering concrete solutions to problems for which the old order could not account, much less provide solutions, any

---

28. This result corresponds to what Spivak calls "worlding" (1985: 243).

29. I refer here to an ideal type in Max Weber's sense, that is, a simplified account of what actually are "complex, contingent processes" (Nabers 2015: 125). This corresponds to what more conventional accounts would call a "model" (King et al. 1994: 49).

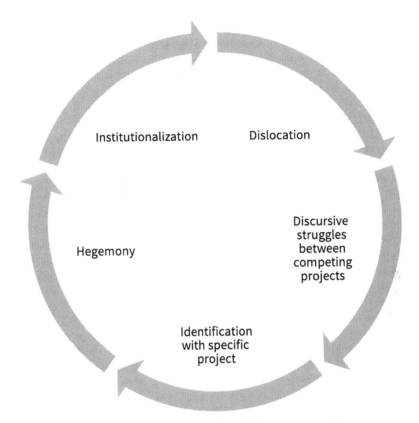

Fig. 2. Hegemonization. (Based on Stengel and Nabers 2019: 256.)

new project has to entail the much broader (and ultimately unavoidably empty) promise that it will realize the vision of a fully constituted identity.

Any hegemonic project has to include an ideological or mythical element, functioning as an "incarnation of the form of fullness as such" (Laclau 1990a: 66), thus concealing the "radical contingency of social relations" (Glynos and Howarth 2007: 113). Such myths provide a powerful vision for a better future, often simultaneously creating a mythical past where the Self was allegedly still whole. Myths essentially provide hegemonic projects with an element of "re-enchantment" (Nesbitt-Larking and McAuley 2017), increasing the chance of a subject's affective investment with the project. This important "fantasmatic" aspect of hegemonic projects has been highlighted particularly by scholars influenced by Lacanian psychoanalysis (Eberle 2019; Solomon 2014; Zevnik 2016). One example of such use of myth is Donald Trump's hegemonic project to "make America great again," which strives to restore the US to its (mythical) Cold War status of being respected in the world (Nabers and Stengel 2019b).

In addition, like individual articulations, larger hegemonic projects have to be credible. Although hegemonic struggles take place under the condition of a dislocated structure, they are not completely unstructured either. Not just any interpretation has an equal chance of being accepted. A discourse approach does not mean that anything goes. Hegemonic struggles always take place on a "partly sedimented terrain" (Thomassen 2005: 295).

Beyond radical discontinuity and credibility, the relative effectiveness of competing hegemonic projects is influenced mainly by three core elements: (1) the articulation of a chain of equivalent demands, (2) the construction of an antagonistic frontier, and (3) representation by which one demand symbolizes the chain of equivalent demands as a whole. These three "ingredients" are intertwined in practice and are only separated here for analytical reasons. They need not all be present. However, as I explain below, articulating a project in such a way that it includes these three features will likely improve the chance that subjects will identify with it.

To begin with, replacing a discursive order—say, a national security order or an entire economic order—requires the forging of a broad unified movement. To achieve this, a hegemonic project has to incorporate a wide range of demands, some of them originally articulated by other groups in society, to increase the chance that a broad number of subjects will identify with it. A project has a better chance of becoming hegemonic if it manages to articulate previously disparate or even contradictory demands as equivalent and going hand in hand, thus forming a "chain of equivalences" (Laclau 1996d) or "equivalential chain" (Laclau 2012: 413). Put simply, the construction of equivalence is an appeal to social actors that "(actually) we want the same [thing]" (Wullweber 2012: 9). Hegemony is essentially about the formation of a unified project out of formerly disparate demands, through which a new discursive order is formed (Howarth 2000: 109). The chain of equivalence includes not only demands in a narrow sense (for, say, external security, environmental protection, development, and peace) but equally different subjects and their (ascribed) demands (e.g., an articulation of "the West" implicitly presumes a convergence of interests among, say, the US, the United Kingdom, France, and Germany).

In principle, a broader chain of equivalent demands—incorporating more disparate or contradictory demands as equivalent—improves a project's chance to become accepted, simply because more subjects will find "their" demands taken seriously and given voice. A project is particularly powerful if it manages to (credibly) articulate previously contradictory demands as equivalent (e.g., the welfare state proclaims to offer a way out of the class conflict between the proletariat and the bourgeoisie). If these demands can be rearticulated as actually equivalent, a hegemonic project can claim to represent not just special interests but the will of "the people" as such.

The breadth of a chain of equivalences is somewhat of a double-edged sword, however. The rearticulation of previously contradictory demands also increases the chance that their rearticulation as equivalent is not credible in light of relevant sedimented practices (Laclau 2014b: 19–20), that is, with what is commonly held to be (factually) true and (morally) right. For instance, the chance that convinced Marxists will identify with a project that articulates proletarian and bourgeois demands as equivalent is slim, simply because accepting it would require abandoning Marxist axioms altogether (Stengel 2019b: 304). Thus, opponents might challenge articulations of equivalence, stressing the particular content of and differences between certain discursive elements instead of their (alleged) equivalence.

This difficulty highlights a more general aspect of hegemony: that the process of the transformation of disparate demands into equivalential ones, of discursive elements into moments of a common discourse, is never complete. As noted above, difference is the basic condition of all social relations, and any moment in an equivalential chain will always be split between an equivalential content and a particular one (Laclau and Mouffe 2001: 113). This is only logical, because if the transformation into moments were complete, individual elements of the chain would be indistinguishable and collapse into one (Laclau 1997: 320–21). As a consequence, "all identity is constructed within this tension between the equivalential and the differential logics" (Laclau 2005a: 70). The totality that emerges as a result of the reduction of different elements into moments within a chain "is, at the same time, impossible and necessary" (Laclau 2006: 107). It is a "failed totality" (Laclau 2005a: 70), because it cannot be complete. Nevertheless, it is unavoidable if any meaning is ever to be fixed. Equivalence and difference are always both present; only the relative balance between the two changes. Moreover, equivalence and difference will often be highly contested. While the logic of equivalence (offensive with respect to the old, disrupted order) strives to construct a common project out of formerly disparate demands, the (defensive) logic of difference stresses the particular content of individual demands, thus undermining the unity of the project (Laclau and Mouffe 2001: 127–34; Nonhoff 2006: 212).

The construction of a chain of equivalent demands is also the point at which representations of reality are linked to (demands for) policy action.[30] Decision makers will usually identify a specific problem and call for a specific set of measures considered adequate to address the problem at hand, incorporating into the emerging chain both policy goals and the deployment of certain instruments as equivalent demands. Consider the example of securitization. One big question in securitization theory remains exactly how threat constructions (i.e., demands to accept the claim that something is a threat) and (demands for) extraordinary means—"emergency measures" (Buzan et al. 1998: 24–25)—are linked (McDonald 2012). From a discourse theoretical perspective, this link happens through their articulation as equivalent demands within the same equivalential chain (Stengel 2019b). For extraordinary measures to become accepted, demands for these measures—surveillance, military operations, and so on—have to be (credibly) incorporated into the chain of equivalent demands. As the discussions in chapters 3–5 demonstrate, this can be a highly complex and contradictory process.

As noted second in the preceding list of aspects influencing project effectiveness, the process of transforming discursive elements into moments is intertwined with the construction of social antagonism, because equivalence between previously disparate or even contradictory demands emerges only as a result of such opposition. Antagonism is the construction of a particular type of Self-Other relationship in which the "radical" (antagonistic) Other is blamed as a "scapegoat" (Solomon 2014: 40) for (1) the fact that certain demands (e.g., for social welfare, peace, and/or security) remain unfulfilled and (2) the incompleteness of the Self's identity (Laclau 1992: 123). Figure 3 illustrates this process. There, D_1 through D_n represent originally disparate or contradictory demands, all of which are unfulfilled. Through the construction of a radical Other, these demands become constructed as equivalent.

These demands are equivalent only in reference to the excluded Other that is said to stand in the way of their realization (Laclau 2005a: 70). Linked to this limitation is the claim that if only the radical Other could be over-

---

30. I thank Martin Nonhoff for reminding me of this point.

come, all demands would be realized. Because all of the previously disparate demands and identities are blocked by the radical Other, they are (articulated as) united in their common opposition to the radical Other that they all need to overcome in order to be fulfilled and/or complete (Laclau 2005a: 70). Thus, the radical Other is reduced to "pure negativity," to a "pure threat" (Laclau 1996d: 38). While the various demands inside the formation are articulated as equivalent, the excluded elements are equally articulated as equivalent, by jointly presenting a threat to the system (Laclau 1996d: 39).[31] Resulting from this process are two opposing chains of equivalences (see fig. 3).

It is important not to confuse social antagonism with other types of Self/Other relationships. Although a detailed discussion is beyond the limited scope of this chapter, the stress poststructuralist studies in IR have put on the importance of an Other for identity formation has sometimes come at the expense of differentiating between different types of Self/Other relationships (critical, Lebow 2008). Thus, many studies (not without reason) emphasize different concepts like difference (Inayatullah and Blaney 2004), danger and threat (Campbell 1998; Rothe 2015), enmity (Behnke 2013; Herschinger 2011), and othering/otherness (Diez 2004; Hansen 2006; Prozorov 2011; Solomon 2014; Suzuki 2007; Yennie Lindgren and Lindgren 2017), not always clearly distinguishing between them.[32]

That approach is not necessarily always problematic, particularly if the purpose is to question essentialist conceptions of identity. However, if we want to understand how identity formation works and how it is linked to related phenomena like threat construction, careful differentiation is indispensable (see Nonhoff 2017a; Rumelili 2015). Moreover, it is equally important to note that while the binary between the Self and the radical Other is crucial for the formation of a hegemonic project (by clearly naming who or what needs to be overcome), the reality of identity formation is infinitely more complex, contingent, contradictory, and multifaceted. For instance, the German security discourse contains references not only to a single radical Other but to various other Others, including allies within NATO or the EU, which are allegedly more militaristic or less enlightened than Germany.

---

31. I am not talking about a physical threat here, in the way threats are commonly understood in IR.

32. This criticism certainly does not apply to all IR poststructuralist studies, some of whom have explicitly discussed different types of Self/Other relationships (Doty 1996; Hansen 2006).

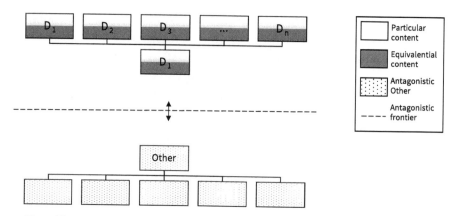

Fig. 3. The construction of a chain of equivalences. (Based on Laclau 2005a: 148.)

Such references are clearly visible in, for example, the debates about the 2003 Iraq War, during which the SPD/Green coalition government positioned Germany as the true advocate of peace (Eberle 2019; Nonhoff and Stengel 2014).

For the purpose of the present study, I understand social antagonism exclusively as that type of Self/Other relationship in which the (radical) Other is articulated as blocking the Self's identity. Antagonism can equally involve the simultaneous construction of the Other as a physical security threat (securitization), as an enemy (enmity),[33] or as morally, intellectually, or otherwise inferior (othering: see Krumer-Nevo and Sidi 2012; Neumann 1996), but it does not need to (see Stengel 2019b). Moreover, if the aim of an analysis is to explain hegemonization, what matters for the forming of a unified collective subject out of previously divided social groups is the construction of an Other as the sole reason why demands remain unfulfilled and identities incomplete; it does not necessarily matter that the Other is inferior, hostile, or physically threatening. In fact, security threats, enmity, and superiority/inferiority are best understood as specific contingent articulations that emerge as a result of discursive hegemony. Thus, from a discourse theoretical perspective, social antagonism is prior to and more fundamental

---

33. See, most notably, Mouffe's (2005a) discussion of antagonism versus agonism in that regard. Similarly, Hansen (2006: 36) has pointed to the possibility of constructing the Self in opposition to a "less-than-radical" Other. Of course, Hansen's understanding of the "radical" Other as an enemy differs from a Laclauian perspective.

than other types of Self/Other relationships. At the same time, as noted above, hegemonic projects can draw on sedimented constructions of Self and Other for credibility, such as when contemporary development or interventionist discourses draw on a familiar hierarchical relationship between a "modern" West and a "traditional" non-West (Chakrabarty 2000; Chandra 2013).

Through the exclusion of a radical Other and the formation of a chain of equivalent demands, a new collective identity is formed. Out of formerly disparate social groups, a common group or movement is forged, united in its quest to overcome the radical Other (Laclau 2014f: 85). In the poststructuralist literature in IR, that a Self can only be constituted by an external threatening Other has been discussed using the term *difference* and/or the term *othering* (Campbell 1998; Inayatullah and Blaney 2004; Walker 1993). Thus, admittedly simplifying things a bit, "the West" emerges as a specific identity only through boundary-drawing practices that delineate it from "the East,"[34] "the rest" (Hall 1992), or various other Others (Behnke 2013; Hansen 2006); "the Occident" only gains meaning by being different from "the Orient" (Said 1979);[35] the identity of the FRG is constituted by its own past (Zehfuss 2002); "our" being "civilized" is constituted by the construction of a "barbarian" Other (Salter 2002); and the world of international politics continues to be mainly influenced by "masculine" virtues because virtues associated with femininity are usually excluded as improper for statesmen acting in an anarchical self-help world (Hooper 2001; Tickner 1988, 1992). Discourse theory shares this emphasis on relationality and difference with other critical approaches in IR, most notably feminist and postcolonial perspectives that highlight how (gendered and racialized) identity constructions usually rest on the exclusion of various Others that are constructed as evil, weak, irrational, backward, or otherwise inferior, thus often legitimizing violent and/or patronizing policies toward them (e.g, Inayatullah 2014; Masters 2009; True and Hewitt 2018; Zalewski 2019). Because of this common emphasis on relationality, a discourse theoretical approach is principally compatible with (many) feminist and postcolonial perspectives. Given

---

34. "The East" is, of course, not a fixed category. Who belongs to "the East," "the West," or some other camp is a question of specific, contingent, and temporary articulation within particular discourses (see Campbell 1998; Klein 1990; Neumann 1999). Thus, from a postcolonial perspective, the Soviet Union might well be perceived as part of the (imperialist) global North rather than "the East."

35. Said's *Orientalism* has received much criticism, including the charge of itself being orientalist (see Huggan 2005).

this compatibility, there is no reason (aside from practical considerations perhaps) why at least a limited investigation of gender and coloniality cannot be woven into an analysis of discursive hegemony. Given the prominent role that gender and coloniality play in the legitimation of "Western" interventions, any analysis that fails to take this role into account (in however limited a fashion) opens itself up to the charge of missing a big part of the picture.

Important to note in this context is that social antagonism is itself ideological, in at least two respects. To begin with, while the radical Other is blamed for the Self's inability to reach a full identity (a perfect society in which all demands are fulfilled), the incompleteness of the Self's identity—indeed, of all identities—is actually due to the chronic dislocation of all social structures. Put simply, it is ontologically impossible to ever reach a full identity. At the same time, meaning can only be fixed temporarily through the reduction of discursive elements to moments of a specific discourse, that is, through the exclusion of all other potential meanings. In this ontological understanding (ontology referring to the nature of being as such), antagonism refers to the "original lack" (Camargo 2013: 174) of dislocation as a general condition of the social (see Laclau 1996d). Meaning can only ever be fixed at the price of incompleteness and exclusion (Laclau 1996d: 37). What happens in the construction of an antagonistic frontier as part of a specific hegemonic project—that is, at the ontic level (the level of concrete discursive struggles)—is that a radical Other is blamed for an identity incompleteness that can actually never be escaped. Conceiving of a unified, pure Self is only possible because of the exclusion of the radical Other. Thomassen aptly observes, "We do not start with a pure inside; the inside is always already dislocated, and it is only the 'negation' of this dislocation—its externalisation—that creates the purity of the inside" (2005: 298). The Other, then, is not just threatening but constitutive, making it possible to even conceive of a coherent Self. If the Other would vanish, the identity of the equivalential chain would be in jeopardy, because it is only held together by the common opposition to the radical Other. As Laclau put it, "any victory against the system also destabilizes the identity of the victorious force" (1996c: 27).

Antagonism is also ideological because the claim that the world consists of two opposing camps only works if the existence of heterogeneous elements (i.e., elements that do not fit neatly into either camp) is denied (Thomassen 2005: 298-99). For instance, in the construction of a chain of equivalence, some particular identities will resist incorporation, because their particular meaning might clash with the particular meaning of other moments in the chain. These elements might not be part of the Other either, which means they cannot be represented and are simply ignored (Laclau 2005a: 139). One example is the Non-Aligned Movement (NAM) during the Cold War. The NAM was a group of states that allied with neither NATO nor the Warsaw Pact (see Miskovic et al. 2014), thus with neither 'the West' nor 'the East.' Only by ignoring the NAM altogether can the story of a bipolar block confrontation be upheld.

The construction of equivalence can often go hand in hand with what Nonhoff (2006: 230–31, 2019: 80) has called "super-differential demarcation" (*superdifferenzielle Grenzziehung*). The drawing of a super-differential boundary involves the banishment of certain elements from a specific discourse, not as antagonistic, but as not thematically belonging to it. In many cases, the boundaries between different thematic discourses are so sedimented that their location (although, to some extent, being principally contested and in flux) is more or less clear. For instance, it is commonly taken as self-evident that, unless one makes the effort of articulating a connection, a lawnmower is not part of a medical discourse (leaving aside lawn-mowing accidents as an area of overlap). Lines are not always this clear-cut, so establishing such boundaries in the first place is often part of a hegemonic project (Nonhoff 2006: 231).

Equally, under certain circumstances, the deliberate breach or relocation of a super-differential boundary can strengthen (or weaken) a specific hegemonic project, by incorporating demands that were previously considered outside of a specific discourse. A case in point is securitization, through which issues previously considered outside the security discourse, like environmental protection, are constructed as matters of security (see Buzan et al. 1998; Rothe 2015). Breaching the super-differential boundary between security discourses and other discourses seems to be an important factor in how securitization processes increase their appeal, by articulating previously unconnected demands or whole discourses as equivalent to security.

The third element previously mentioned as increasing a hegemonic project's chance of success is representation. Hegemonic discourses are organized around (sets of) privileged signifiers, called "nodal points" or *points de capiton* (Laclau and Mouffe 2001: 112), which function as anchors. Among these priviledged signifiers that are particularly important for the functioning of a discourse, the "empty" or "master signifier" is crucial (Laclau 1996d:

36; 2006: 107). Essentially, one particular demand that is part of the chain of equivalences (D_1 in fig. 3) assumes the function of representing the chain as a whole. This is called a "hegemonic relationship" (Laclau 1996d: 42). This specific signifier empties itself of its particular content, thus becoming a "signifier without a signified" (Laclau 1996d: 36). By detaching itself from its particular content, an empty signifier can function as a symbol with which subjects can identify and to which they can connect their particular demands. In this way, it functions as "a surface for inscription able to register a series of demands and interests much broader than its initial form of articulation" (Howarth 2000: 173). For instance, the signifiers "freedom" and "justice" have such broad meanings in different contexts that a wide-ranging number of demands can be voiced in association with them. Ideal-typically, the hegemonic relationship is mirrored on the other side of the antagonistic frontier, with one moment in the chain symbolizing the radical Other to be overcome. Through this operation, the discursive formation receives a name and is constituted as a unified object. Thus, instead of, say, "justice" having an a priori essence, what the signifier means is determined in the moment of the creation of a hegemonic relationship. As Laclau put it, "the name is the ground of the thing" (2006: 109, italics removed). The empty signifier assumes the function of a temporary center of the discursive formation (Laclau and Mouffe 2001: 112).

Importantly, the empty signifier not only represents the chain of equivalences but becomes "the symbol of a missing fullness" (Laclau 1996c: 28). It symbolically stands in for a larger vision of a "perfect society" (Herschinger 2012: 86) in which all demands are fulfilled and in which a complete, unadulterated, undisrupted identity is reached. Thus, the empty signifier also symbolizes the universal. Again, this operation is unavoidably ideological, because a fully constituted identity, a perfect society, is ontologically impossible (Laclau 2014b: 17–18). The same phenomenon usually happens on the flip side of the antagonistic frontier, with one particular identity symbolically representing not just the antagonistic chain of equivalences as a whole but the constitutive lack at the heart of all social relations.

Although previous studies have focused on the importance of a single empty signifier like "freedom" (Nabers 2015), "the people" (Laclau 2005a), "international community" (Herschinger 2012: 86), "nanotechnology" (Wullweber 2014: 289–92), or the "social market economy" (Nonhoff 2006), I follow Reyes (2005) in his argument that the function of an empty signifier can be fulfilled by a number of different signifiers that refer roughly to the same social demand. Reyes refers to this as "over-wording" (109). As I demonstrate in the following chapters, this fulfillment occurs in the project of networked security, where a number of different signifiers ("comprehensive security," "networked security," etc.) assume the representation of the overall discourse. I refer to such a (set of) signifiers as a "blurred empty signifier," in the sense that its representative function is assumed by a number of similar signifiers, making it appear somewhat blurred.[36]

Two further aspects are worth noting with respect to representation. First, which signifier comes to represent the overall chain is not determined by the (dislocated) structure but is the result of discursive struggles. As a consequence, prior to an empirical analysis, there is no way of determining which difference will be crucial (Laclau 1996d: 43). Second, the empty signifier remains "split between its equivalential content and its differential content" and thus "only tendentially empty" (Laclau 2006: 107; Thomassen 2005: 293). Because the empty signifier retains a "minimal remainder" (Howarth 2000: 39) of its particularity, which signifier assumes a privileged position makes a significant difference, because the signifier in that position influences the meaning of the overall discourse. For instance, as is nowhere more apparent than in processes of securitization, it makes quite a difference whether a discourse on immigration is constructed around "national security" or, say, a "duty of care."[37] At the same time, because the tension between particularity and equivalence can never be resolved, the hegemonic relationship remains unstable and somewhat contradictory (Laclau 2006: 108). That situation is further complicated by the special role of the empty signifier that is both another differential-cum-equivalential element of the chain and somehow "above" the others, as it comes to represent all of them (Nonhoff 2006: 218). For example, understanding climate change as a security issue might mobilize support for measures to counter the problem (although not necessarily so: see McDonald 2012), but it simultaneously opens this particular representation up to criticism that understanding climate change this way is misleading or even counterproductive (e.g., Mason and Zeitoun 2013), thus potentially destabilizing the discourse.

---

36. The term *blurred empty signifier* was proposed by Martin Nonhoff during a personal conversation with the author.

37. I here refer to care ethics as proposed in feminist scholarship, emphasizing altruism and a responsibility toward the Other (Hutchings 1994: 28–29).

CONCLUSION

In this chapter, I have developed a theoretical framework with which to analyze social change as the result of discursive struggles and hegemonic processes, drawing attention to the crucial notions of sedimented practices,
equivalence, social antagonism, and representation. Discourse theory offers
a comprehensive framework with which to explain the relative effectiveness
of different articulations and discourses. Given discourse theory's far-
reaching claim that the social is coextensive with the discursive, the framework outlined here should have a wide applicability to phenomena of interest to IR theorists, including, for instance, norm dynamics (Renner 2013),
securitization processes (Rothe 2015), or the emergence and demise of international orders (see Stengel and Nabers 2019). Moreover, discourse theory's
social ontology allows for it to be combined with arguments from feminist
and postcolonial theories, the poststructuralist variants of which equally
stress discourse as the central terrain in which gender, race, and modernity/
coloniality are (re)produced (e.g., Butler 1990, 1993; Said 1979). The next
chapter addresses the consequences that adopting a discursive ontology has
for research practice and how to go about a poststructuralist discourse
analysis.

CHAPTER 2

Hegemony Analysis as Reconstructive Social Research

Methodology and Methods

Building on the preceding outline of the general theoretical framework of discourse theory, this chapter discusses how to do empirical research based on a discursive ontology of the social. The chapter is structured in three main sections. The first examines the metatheoretical implications of accepting discourse theory's ontological framework. It locates discourse theory in terms of its ontological presuppositions and discusses its consequences for epistemology and metaethics.[1] In contrast to claims that question the utility of metatheoretical debate (Halliday 1996: 320; Rosenau 1996: 313), this chapter shares Marsh and Furlong's observation that there is no escape from metatheoretical commitments, implicit or explicit, which unavoidably influence one's research process (Marsh and Furlong 2002; also Hammersley 1992). Thus, the best way to address this potential problem is to face the issue head-on.

A metatheoretical clarification seems particularly sensible with respect to poststructuralist approaches, such as the one advanced here. Despite the publication of a number of works that explicitly address these issues (e.g., Hansen 2006; Nabers 2015), the precise metatheoretical position of different poststructuralist approaches remains often only poorly understood, giving rise to criticism that is at least not always fully justified. Moreover, although this book will likely appeal mostly to readers familiar with poststructuralist

---

1. In that sense, I share the presupposition that epistemological assumptions follow from one's ontological framework. See, e.g., Wight 2007b: 385.

and, more generally, critical approaches in IR, it tries to engage with a general IR audience, including scholars skeptical of poststructuralist approaches. Mainly for their benefit, this chapter's first section addresses metatheoretical criticism in detail, at the risk that some of the discussion might seem repetitive to readers already familiar with poststructuralism (who are welcome to jump directly to the discussion of research design or to the more specific section on methods). Poststructuralism is a broad tent, filled mostly with people rejecting the label (see Angermüller 2015), and my discussion here is limited to the position advanced by Laclau and Mouffe. Given, however, that a number of leading IR poststructuralist thinkers draw on Laclau and Mouffe's ontological framework (e.g., Campbell 1998; Doty 1996; Edkins 1999; Hansen 2006; Nabers 2015), the discussion here equally applies to a significant part of that field of research.

The second section of this chapter turns to the more practical aspects of doing poststructuralist discourse analysis, addressing questions of research design, what it means to provide a discourse theoretical explanation (an understanding that differs significantly from a neopositivist explanation based in political science: see, e.g., King et al. 1994; Van Evera 1997: 15), and what role politics and normative commitments play in research. I locate poststructuralist discourse theory broadly within what is often called "qualitative" (Denzin and Lincoln 2003a, 2003b, 2005b), "reconstructive" (Bohnsack 2014; Franke and Roos 2013b; Herborth 2010), or, more specifically, "interpretive" research (Schwartz-Shea and Yanow 2012; Wagenaar 2011; Yanow 2009; Yanow and Schwartz-Shea 2006a). This chapter's third section turns to the question of corpus selection and how the abstract theoretical concepts of discourse theory can be translated into more specific categories suitable for discourse analysis.

ON THE METATHEORETICAL LOCATION OF DISCOURSE THEORY

Questions of metatheory or "meta-methodology" (Bevir 2008) continue to be hotly debated in IR,[2] including the precise concerns of different forms of "second order" theorizing (Wendt 1991: 383) and where to place even prominent positions like (neo)positivism (Bates and Jenkins 2007). In this section,

---

2. See, e.g., the contributions to the first issue of *International Studies Review* 17 (2015) and the third issue of *Millennium: Journal of International Studies* 43 (2015).

I briefly clarify what is meant in the context of this study by references to ontology, epistemology and metaethics. I address methodology, sometimes counted as part of metatheory (e.g., Mayer 2003), in the subsequent section.

I here understand ontology as concerned with the nature of being (*Seinslehre*), that is, "the study of what there is" (Hofweber 2014; see Monteiro and Ruby 2009: 25).[3] Patomäki and Wight (2000: 215) further distinguish between "philosophical" and "scientific" ontology. Philosophical ontology focuses on what Jackson has called researchers' "'hook-up' to the world," that is, the (very general) question of whether there is a world "out there" that is independent of our "knowledge-making practices" (2010: 28, 31). Scientific ontology is concerned with the specific setup of the world, that is, what entities exist (P. T. Jackson 2010: 28)—say, states, individuals, discourses, social structures—and what their properties are (e.g., if people are best understood as "walking wave functions," per Wendt 2015: 3).[4] Epistemology asks about our ability to produce knowledge (*Erkenntnistheorie*), that is, the "knowability" of things (Yanow and Schwartz-Shea 2006b: xi; Marsh and Furlong 2002: 18–19; Steup 2014). Closely related to epistemology is metaethics, which focuses on the possibility to formulate normative or moral statements based on universal principles (Reus-Smit 2012; Sayre-McCord 2012; Singer 1993: ch.

---

3. Ontology equally includes debates about how to settle questions regarding what exists (or not) (Hofweber 2014), which is, strictly speaking, a matter of how to gain knowledge and, as such, makes it difficult to neatly distinguish ontology from epistemology.

4. Here, I use the distinction between philosophical and scientific ontology mainly for practical reasons, although it is not unproblematic if situated within philosophical debates. Not only is ontology contested (Hofweber 2014), but it is doubtful whether scientific ontology, as it is understood here, would even fall within the field of ontology as it is understood in philosophy. Ontology emerged as a distinct philosophical term and a subdiscipline of metaphysics (which asks about the "things" in the world and their characteristics) only in the 17th century, around the time when phenomena like freedom of will or the relationship of mind and body, previously considered to belong to other branches of philosophy (most notably physics), came to be discussed within metaphysics, as the "study of being" is called in philosophy. These questions were previously considered to concern not the nature of being as such but specific entities (van Inwagen and Sullivan 2015). Consequently, one would have to ask whether the study of the being of, say, states is at all a matter of ontology, philosophically speaking, because we can legitimately ask whether such study tells us anything about being as such. But the borders between ontology and metaphysics more generally are "a little fuzzy" anyway (Hofweber 2014). Since the present book is primarily a substantive study in IR, not a contribution to metatheoretical debates in IR, I leave the complications of defining ontology at that and wish the reader good fortunes at pursuing it further, if so desired.

1). To the extent that the fact/value distinction is considered problematic, epistemology and metaethics merge, as truth claims are at least implicitly normative (Cox 1981; Horkheimer 1937; see also P. T. Jackson 2010). For analytical reasons, I keep epistemology and metaethics separate here. Table 1 provides a brief overview of different metatheoretical branches and of criticism leveled against poststructuralist approaches.

As just noted, philosophical ontology is concerned with the broad question of whether reality has an existence independent of our representations about it. The default position in terms of ontology is philosophical realism (in philosophy, simply "realism"), which presumes the existence of a world independent of our knowledge-making practices.[5] Diametrically opposed to this position is philosophical idealism (i.e., metaphysical or ontological idealism, as it is called in philosophy: see Guyer and Horstmann 2015), which refers to the ontological assumption that "something mental . . . is the ultimate foundation of all reality, or even exhaustive of reality" (Guyer and Horstmann 2015). Seen this way, "the world" and our descriptions of it are "one and the same thing," and, consequently, "changes in description lead to changes in the object itself" (Joseph 2002: 121, 12).

There continues to be some confusion about exactly where poststructuralism comes down on the issue of philosophical idealism, and a number of statements by poststructuralist (as well as pragmatist) thinkers lend

TABLE 1. Metatheoretical criticism of poststructuralism

Metatheoretical branch	Subject matter	Criticism leveled against poststructuralist approaches
Ontology	The nature of being and how to determine what exists in the world	Philosophical/ontological idealism
Epistemology	The possibility of factual/epistemic statements	Epistemic relativism
Metaethics	The possibility of moral/normative statements	Moral relativism

---

5. Jackson discusses this position under the heading of "mind-world dualism" (P. T. Jackson 2010: 31, italics removed). Leaving aside the question of whether a term including *mind* is appropriate for discursively produced knowledge (a question meriting a discussion in its own right; see Nabers 2015, 2018), Jackson helpfully points out that a correspondence theory of truth rests on philosophical realism because it only makes sense to discuss a gap between our knowledge and the "real world" (i.e., epistemological matters) if the two are distinct.

themselves to misinterpretation. Examples include Laclau's claims that "discourse is the primary terrain of the constitution of *objectivity as such*" (2005a: 68, italics added) and that "natural facts are also discursive facts" (Laclau and Mouffe 1987: 84), the insistence by Howarth and Stavrakakis that "*all* objects are objects of discourse" (2000: 3, italics in original), Butler's assertion that discourse "bring[s] about what it names" (1993: 13), the proclamation of the "death of the author" by Barthes (1967), Campbell's argument that "nothing exists outside of discourse" (2001a: 444), or Kratochwil's claim that "the objects of experience are not simply 'there' in the outer world but are the results of our constructions and interests" (2007b: 6). Despite what these statements seem to suggest, they are not advocating philosophical idealism. Poststructuralists claim not that physical, material objects change with our representations of them (discursive or mental) but only that the meaning (or identity) of an object, subject, or practice is discursively produced. As Laclau and Mouffe argue, "the fact that every object is constituted as an object of discourse has *nothing to do* with whether there is a world external to thought" (2001: 108, italics in original). Poststructuralists doubt "not that such objects [furniture, stones and all that hard stuff of the material world] exist externally to thought [they do]" but "the rather different assertion that they could constitute themselves as [intelligible] objects outside any discursive condition of emergence" (Laclau and Mouffe 2001: 108).

In a different context, Laclau and Mouffe use the example of a stone to illustrate this claim.

> Again, this [the discursive construction of reality] does not put into question the fact that this entity which we call stone exists, in the sense of being present here and now, independently of my will; nevertheless, the fact of its being a stone depends on a way of classifying objects that is historical and contingent. If there were no human beings on earth, those objects that we call stones would be there nonetheless; but they would not be "stones," because there would be neither mineralogy nor a language capable of classifying them and distinguishing them from other objects. (Laclau and Mouffe 1987: 84)[6]

---

6. Even Wendt—not suspicious of any poststructuralist fits (yet: see Wendt and Duvall 2008)—acknowledged that regarding meaning as produced by discourse "does not require a denial of reality 'out there.' . . . The claim is merely that reality has nothing to do with the determination of meaning and truth, which are gov-

A stone's physical existence does not change just because we proclaim it to be a tree, but how we understand it (as a convenient "natural furniture" to sit on, a decorative item to take home, a research object to study, or a weapon) depends on how it is endowed with meaning in discourses. This is precisely the position of Wendt (hardly known for being a philosophical idealist) when he points out that "material power is only 'power' insofar as it is meaningful" (2006: 213).

A different example clarifies how this understanding relates to social analysis. In the (admittedly somewhat macabre) event of a person getting hit in the head by a bullet (see Grint and Woolgar 1992), scientific realists (e.g., Wight 2007b) and poststructuralists, for example, agree that this person will die if not admitted to a hospital very quickly. Death—or, rather, the biological processes that have been named "death"[7]—is simply the likely consequence of having a bullet lodged in your brain, and not even the most ardent poststructuralist would doubt that. Where poststructuralists depart from critical or scientific realists is that for the former, as Grint and Woolgar explain, the meaning of the incident, its existence as death (and not something else), counts.

> If the victim was dead or unconscious or unaware of the dangers, then he or she would not be able to account for the injury. So only those victims still conscious or the shooter or those just watching would be able to verify what happened. Only those people familiar with guns would assume a relationship between bullet and human reaction; only those certain that the victim was not an actor in a film set would be fairly sure that the wound was "real"; only those actually present could verify the connection—the hospital surgeon retrieving the bullet could not guarantee that the wound was caused by the bullet. In short, most people would have to take it on trust that the wound was caused by the bullet. (Grint and Woolgar 1992: 376)

As far as the social consequences are concerned, what matters is the event's discursive articulation, not the pure physical fact of some material object (the bullet) being located inside a different material object (a person's head). The same incident could assume the meaning of an accident, suicide,

---

erned instead by power relations and other sociological factors within discourse" (Wendt 1999: 55).

7. Interpretation does, of course, depend on one's definition of death (Grint and Woolgar 1992: 376).

murder, a tragedy, or (in a war) a lawful killing, depending on the dominant interpretation of the issue (Kessler and Werner 2008: 263; Zehfuss 2018). That interpretation makes certain social consequences more or less probable. A murder would, for instance, most likely result in a trial (i.e., if everything goes right), whereas a war death would not (necessarily). Likewise, the victim's spouse might benefit from life insurance in the case of an accident but perhaps not in the case of suicide. The physical event itself (person hit by bullet) remains the same, but the meaning and, by extension, its social consequences are completely different in various contexts. Undoubtedly, a physical, material world exists independently of the observer's mind (or discursive representations, which should not be conflated: see Nabers 2018), but without reference to discourse, it simply has no meaning to us. This is why poststructuralists argue for meaning to assume center stage in the social sciences.

Having established that discourse theory is not ontologically idealist does not tell us much about discourse theory's actual position. Here, it is helpful to further distinguish between realism and various types and degrees of nonrealism, differentiated according to two dimensions. First, a realist position can be distinguished from nonrealism with respect to existence, whether something actually exists or is a mere illusion, and independence, whether something is independent of our representations of it (Miller 2014). Second, positions can vary with respect to the social and the natural (physical) world; people can be realists with respect to the natural world but nonrealists with respect to the social world.

Using this simple heuristic, one can venture an educated guess about where discourse theory can be located in ontological terms (table 2 provides an overview). To begin with, as I have pointed out in the preceding discussion of idealism, discourse theory does not doubt the existence of a natural world independent of our knowledge-making practices. It is ontologically realist regarding the physical, natural world (which, however, is unintelligible by itself).

TABLE 2. The ontological presuppositions of discourse theory

	Aspects of realism	
	Existence	Independence
Social world	Realist	Nonrealist
Natural world	Realist	Realist

Things are more complicated if we turn our attention to the social world. In this regard, I argue, discourse theory is partially nonrealist. On one hand, discourse theory's discursive ontology of the social presumes that the social world exists (philosophical ontology) and that it is made up of discourses (scientific ontology) (Glynos and Howarth 2007: ch. 4; Howarth 2010: 313; Marchart 2007: 146–49). Because the discursive ontology of the social is not an instrumentalist claim (see Monteiro and Ruby 2009) but an ontological one, discourse theory has to be understood as realist with respect to the aspect of existence. On the other hand, in regard to independence, discourse theory is nonrealist, claiming not only that social facts like norms are mediated by discourse but that they are literally produced in discourse (see chapter 1). Thus, although social facts are certainly independent of what individuals think or say about them, they are not independent of collective representations of them. As a consequence, discourse theory sees the social world not as independent of but as deeply intertwined with our knowledge-making practices (our articulations), which is a nonrealist position.

Having clarified discourse theory's ontological position allows us turn to epistemology (truth claims), to metaethics (normative statements), and to the question of relativism. Although the precise meaning of the term *relativism* is subject to debate in philosophy, it broadly refers to a family of views united by the common theme "that some central aspect of experience, thought, evaluation, or even reality is somehow relative to something else" (Swoyer 2010).[8] I distinguish here between strong and weak versions of relativism, as well as between normative or moral relativism and descriptive or epistemic relativism (see table 3).

Descriptive or epistemic relativism refers to the epistemological view that there is no objective, absolute truth. Rather, truth (as well as rationality:

TABLE 3. Examples for different types of relativism

	Descriptive/epistemic (epistemology)	Moral/normative (metaethics)
Strong	Any truth claim is as valid as every other.	There is no basis to distinguish between "right" and "wrong."
Weak	Facts are theory-laden.	What is good or bad behavior in a specific situation depends on socially constructed norms.

---

8. In that sense, ontological idealism could be seen as a form of relativism in that the world's existence would be relative to our conceptions of it.

see Sankey 2010: 1) depends on context (Bunge 1996). Similarly, as a meta-ethical position, normative or moral relativism states that the truth or falsity of normative statements is not universally valid but depends on the normative framework of a particular society or group of people (Gowans 2012). In its strong version, relativism refers to the standpoint that "every belief on a certain topic, or perhaps about *any* topic, is as good as every other" (Rorty 1982: 166, italics in original). Weak relativism refers to the view that not all statements, such as normative ones, are equally accepted as true in every society. Thus, while strong relativism is more controversial and tends to lead to inconsistencies, weak versions are less controversial but can tend to be somewhat trivial (Cilliers 2005; Swoyer 2010). In particular, strong versions of relativism are usually used to charge opponents rather than by authors claiming the label for themselves. As Swoyer has pointed out, "most academic philosophers in the English-speaking world see the label 'relativist' as the kiss of death" (Swoyer 2010), and it indeed seems doubtful that anyone actually holds a strong relativist view (Rorty 1982: 166).

That charges of strong relativism continue to resurface is due to what discourse theorists call "post-foundationalism" (see Howarth 1995: 117; Marchart 2007),[9] an epistemological position that, contrary to empiricism (see Nicholson 1996: 128), holds that there are no ultimate, final foundations on which to ground truth claims. Since (the meaning of) everything is discursively constructed, there are no independent meaningful facts that can function as a neutral arbiter between truth claims (Nonhoff 2011: 92).[10] Since the same "facts" acquire different meaning depending on the discourses (including scientific theories) in which they are articulated, they cannot function as a neutral arbiter between competing truth claims. In that sense, discourses function like colored glasses, unavoidably influencing how we see the world. From a discourse theoretical perspective, it is impossible to look at the world without such glasses: one can substitute one set for another, but one cannot get rid of them. As a consequence, like pragmatists (see Rorty 1979; 1981), discourse theorists argue that there is no universal external independent yardstick against which to measure our knowledge (or normative

---

9. E.g., Geras 1987: 67; Sayer 1993; Sweetman 1999; White 1992: 78. Similar criticism persists among IR scholars: e.g., Mayer 2003: 88; Wight 1999a: 314; 2007a: 41; Worth 2011: 385.

10. This position corresponds to what Monteiro and Ruby (2009: 17) discuss under the heading "social constructivism," namely, "the position that truth is a function of social and political processes," as a result of which objectivity is impossible .

commitments) and establish whether it is right/true or wrong/false, like a correspondence theory of truth would claim (Friedrichs and Kratochwil 2009: 703). Rather than being an objective feature of reality, truth is a matter of discursive construction. As Laclau and Mouffe point out,

> It would be absurd to ask oneself if, outside all scientific theory, atomic structure is the "true being" of matter—the answer will be that atomic theory is a way we have of classifying certain objects, but that these are open to different forms of conceptualization that may emerge in the future. In other words, the "truth," factual or otherwise, about the being of objects is constituted within a theoretical and discursive context, and the idea of a truth outside all context is simply nonsensical. (1987: 85)

The claim that there is no absolute, universal, ahistorical foundation for truth does not mean, however, that there can be no foundation at all. Post-foundationalism, as opposed to anti-foundationalism (see Marchart 2007), does not mean that any truth claim is as good as any other (Kratochwil 2007b: 11). In fact, rather than having no means at all of distinguishing between truth claims, discourse theorists would claim that we have context-dependent, imperfect, and limited knowledge (Cilliers 2005: 160; Marchart 2007: 2, ch. 1).[11] From a discourse theoretical perspective, what makes a statement true or false is not the degree to which it "objectively" matches an independent reality but whether an interpretation makes sense within a specific discursive or theoretical context, because direct access to the "real world" is beyond our grasp. Thus, rather than a correspondence theory of truth, discourse theory would advocate what Friedrichs and Kratochwil call a "consensus theory of knowledge" (2009: 710), all the while acknowledging that neither scholarly discourse is a level playing field. In line with a pragmatist stance, the acceptance or rejection of a truth claim depends on the established standards in a specific scholarly community, which involves interpretation (agreement) by the audience (Kratochwil 2007a: 59–60), influenced by the scholarly standards established in that discipline (see also Winch 1990), which themselves are understood here as sedimented practices.

The same goes for normative statements (metaethics). While post-foundationalism certainly questions the validity of universal grounds for

---

11. Whether anyone is actually a true foundationalist is questionable, in that almost no one today really assumes, in a strict naive empiricist sense, that we have unmediated access to reality (Abbott 1990; Bucher 2017; Skinner 2002).

ethics, this questioning does not mean that any moral statement is equally valid if compared to any other. After all, norms (e.g., whether it is acceptable to eat pigs, cows, dogs, or cats) not only vary in different cultural settings but change over time. The reason for that variation is not that God handed down a new set of commandments or something similar but that the normative framework of society has changed. The validity of any normative statement will depend on the normative framework of a given society (in discourse theoretical terms, sedimented practices), and the latter is a product of past human action. As Laclau points out, "there is no logical transition from an unavoidable ethical moment, in which the fullness of society manifests itself as an empty symbol, to any particular normative order"; rather, the emergence of any particular normative order is the result of an "ethical *investment*" (2000: 81, italics in original).

Nevertheless, denying the possibility of any universal normative foundations does not mean that we need to give up normative commitments altogether (see Mustapha 2013). It just means, notwithstanding any "Cartesian Anxiety" (Bernstein 1983: 16–19) this might cause, that outside of human societies themselves, there is no authority that will make the tough decisions for us. Claiming that moral judgments are context-dependent is not the same thing as arguing that we cannot make a normative commitment. Indeed, it only stresses that since no extra-discursive authority (such as an objective reality that can be accessed directly) decides for us, our decision to identify with specific normative standpoints takes place on an undecidable terrain, which is why responsibility ultimately resides with the subject (Wight 1999a: 314).[12] In that sense, post-foundationalism only points to the (admittedly uncomfortable) fact that we ourselves are ultimately responsible for the normative orders that we support.

METHODOLOGICAL CONSIDERATIONS

How does discourse theory's ontological framework translate into research practice; that is, what are the methodological implications of the ontological, epistemological, and metaethical presuppositions laid out above? I understand methodology here as entailing a translation of abstract ontological and epistemological presuppositions into concrete "rules for scientific

---

12. See the discussions of aporia in Zehfuss 2002 and 2018.

practice" (Mayer 2003: 51). Sometimes called "ways of knowing" or "logics of inquiry" (Haverland and Yanow 2012: 401; Moses and Knutsen 2007), methodology is not to be confused with methods, the instruments used in data gathering and analysis, such as participant observation or surveys (Haverland and Yanow 2012: 401; Schwartz-Shea and Yanow 2012: 4), the choice of which depends not only on the research question (that too, of course) but also on methodological choices as well as the research material analyzed (Haverland and Yanow 2012: 401; for a detailed discussion, see Moses and Knutsen 2007; Schwartz-Shea and Yanow 2012).

Discourse Analysis as Reconstructive Research

Since the central interest of discourse theory is "the way in which political forces and social actors construct meanings within incomplete and undecidable social structures" (Howarth 2000: 129), as well as the social consequences of adopting one particular representation of reality instead of another, an empirical research project employing discourse theory must make use of some form of discourse analysis, which should be understood not as a single unified method but, rather, a "research perspective" (Keller 2013: 3, italics removed). Central for a discourse analysis is the aim to reconstruct the rules and/or regularities according to which meaning is produced in a specific discourse (Keller 2013: 3). Discourse analyses aim at recovering what Foucault (1980: 83) called a "historical knowledge of struggles," thus revealing that truths taken for granted are the contingent product of such struggles. This aim distinguishes discourse analyses from other forms of textual analysis (although discourse analysis does not need to focus on texts in a narrow sense), such as operational code analysis, which seeks to infer actors' beliefs from their statements (e.g. Hermann 2005; Schafer and Walker 2006). Discourse analysis is agnostic regarding whether actors "really mean what they say" (Wæver 2004: 199).[13] While factors like beliefs or norms, prominent in psychological and constructivist approaches, need to be internalized by actors to have an effect on their actions,[14] discourse analysis is interested in how things are understood in the public realm and

---

13. Philosophers even debate whether and how we can know that other people also have inner lives (the so-called problem of other minds: see Hyslop 2015), let alone their mental state.

14. This internalization is problematic from a methodological perspective (see Hanrieder 2011; Hellmann 1999; Krebs and Jackson 2007; Nagel 1974).

(depending on the specific research question) what this means for the realm of acceptable policy options. Exactly how a discourse theoretical approach is implemented in practice depends on the research question, the theoretical framework, and the disciplinary context (Keller 2013: 3) and can range from an in-depth analysis of a small text or visual image (Schlag and Heck 2012) to the use of quantitative methods of textual analysis (Glasze 2007a; Nabers 2015).

The present study relies on a specific form of discourse analysis, so-called hegemony analysis, proposed by Nonhoff (2006, 2007b, 2008, 2019) as a way to "operationalize" the abstract theoretical concepts of discourse theory for empirical analysis. I get back to the technical aspects of coding below. Here, I focus on how abstract theoretical concepts and empirical analysis are brought together in such a form of discourse analysis.

The central purpose of a hegemony analysis is to examine processes of hegemonization. Above all, it seeks to answer the question of how a certain discourse (a certain representation of reality, certain moral standards, etc.) has become dominant. As a consequence, any hegemony analysis starts with the theoretical concepts of discourse theory. In that sense, hegemony analysis could be said to proceed in a broadly deductive fashion (Nonhoff 2008: 301). However, this general movement from theory to empirical material should not be mistaken for a more conventional understanding of deduction as hypothesis testing, for a number of reasons.

First, a discourse theoretical approach that is philosophically rigorous (per Bevir, quoted in Yanow and Schwartz-Shea 2006b: xvi) is not compatible with a strictly deductive approach. In the "deductive-nomological" model of explanation, a specific case is explained through a general law and the explication of antecedent conditions (Hempel 1962), and theories are subjected to empirical tests with the aim of falsification (Popper 2005). Hypothesis testing implies a correspondence theory of truth (Blaikie 2004a: 377; 2004b: 243; David 2013; Hempel 1950; Van Evera 1997: 28), which is incompatible with discourse theory's basic premise that reality is discursively produced.[15]

---

15. Ontological realism and a correspondence theory of truth are not even unproblematic in the hard sciences. One example is Schrödinger's (1935) (undead) cat, a thought experiment in which a cat is kept in a box together with a glass tube filled with (highly poisonous) hydrocyanic acid. A hammer mounted above the glass tube will be released automatically if a Geiger counter registers radiation emanating from an instable atom positioned next to it. If the atom decays, the hammer will break the glass tube, and the cat will die. The problem is that, according to quantum mechanics,

Second, a purely deductive approach that employs predefined categories would risk that the researcher misses an important discursive pattern not specified a priori by the theory (Franke and Weber 2012: 672). For the purpose of explanation (as opposed to a hypothesis test), such an approach is only of limited value. Instead, any discourse analysis has to approach the research material in an open fashion, with a willingness to be "surprised" by the data (Franke and Roos 2010: 285), even if that entails the abandonment of tentative theoretical concepts and the formulation of new hunches or more general statements about potential connections between phenomena and aspects of them. In that sense, a discourse analysis is similar to a grounded theory approach (see Charmaz 2003; Glaser 2002; Glaser and Strauss 2006). A discourse theorist does not approach the research material in a purely inductive, *tabula rasa* fashion either, because that would imply that reality can speak for itself.[16]

Third, rather than following either a purely deductive or a purely inductive approach, discourse analysis involves "jumping back and forth between theory and analysis" (Herschinger 2011: 46). Such an approach combines inductive, deductive, and abductive/retroductive reasoning (Franke and Roos 2013a: 14; see also Glynos and Howarth 2007: ch. 1). It involves a constant back-and-forth between individual texts and abstraction on the level of the discourse as a whole, similar to a hermeneutic circle (Ramberg and Gjesdal 2014).

Fourth, in addition to the primary theoretical framework that guides the analysis, secondary theoretical concepts usually influence it, but without being codified into formal analytical categories prior to beginning analysis. In the case of the present study, this influence mostly concerns theoretical insights from what can broadly be described as critical IR and Critical Secu-

---

the atom remains in a state between decayed and not decayed—a state Schrödinger (1935: 812) calls "mixed" (*vermischt*) or "smudged" (*verschmiert*)—until the measurement takes place (e.g., by lifting the lid of the box), at which point it jumps into one of the two states. As a consequence, the cat must logically exist in a state between being dead and alive, having been poisoned or not as an indirect result of the decay (or not) of the atom (Schrödinger 1935: 812). The state of the cat as being either dead or alive would only be decided through observation, which cannot be neatly separated from the state it is supposed to only describe. As the example illustrates, from a quantum perspective, even the physical states of natural entities (like a cat) are neither clearly determinable nor independent from observation, which challenges ontological realism and a correspondence theory of truth. Not surprisingly, then, Schrödinger criticized what he calls a "naïve realism" (1935: 823).

16. See Waltz 1979: 4, for a similar discussion from a quite different point of view.

rity Studies or Critical Military Studies, as well as critical geopolitics, particularly feminist and postcolonial scholarship. The role these concepts played in the analysis is best understood as like that of sensitizing concepts in Grounded Theory (Bowen 2006: 13–14). As opposed to definitive concepts (see Wonka 2007), sensitizing concepts function as "interpretive devices" that "draw attention to important features of social interaction" and thus help distinguish important from unimportant aspects of the research material (Bowen 2006: 13–14).

As noted at the outset of this chapter, a hegemony analysis closely corresponds to what has been loosely referred to as reconstructive, qualitative, or interpretive research.[17] Reconstructive/qualitative research (the term I will use in the following) is a relatively broad tent. It includes approaches that have different metatheoretical positions, rely on different theoretical frameworks, and use different methods and techniques of data gathering and analysis, such as (auto)ethnography, narrative interviews, participant observation, archival research, grounded theory, or discourse analysis (Richardson 2003).

Much has been written about qualitative research methodology,[18] so a brief overview shall suffice here. In a nutshell (and at the risk of oversimplifying a bit), what unites different qualitative approaches is that they approach their objects of analysis in a similar way that distinguishes them from what Jackson (2010) refers to as neopositivist research in the social sciences. Neopositivist research (I am again simplifying here) tends to embrace ontological realism; relies on a correspondence theory of truth, on theory testing, and on technical rules for case selection with the aim of generalizing findings beyond the examined case(s); strives for objectivity by way of methodological techniques; tries to clearly distinguish between (neutral) facts and (normative) values; and seeks to identify causal relationships between different social phenomena that are usually conceptualized as variables.[19] The

17. Ralf Bohnsack (2014: 11), the most prominent advocate of a reconstructive approach in (German) sociology, refers to "reconstructive" and "qualitative" social research interchangeably. Interpretive research is sometimes equated with qualitative research (Schwartz-Shea and Yanow 2012) and sometimes defined more narrowly, excluding, for example, critical realist approaches (Marsh 2015). I will use the terms here interchangeably, although that is somewhat imprecise, because my main aim is to outline how reconstructive/qualitative/interpretivist approaches differ from neopositivist ones.

18. For overviews, see, e.g., Denzin and Lincoln 2003a, 2003b, 2005b; Flick 2006.

19. Here also, a vast body of research is available: see, e.g., Brady and Collier 2004; Goertz and Mahoney 2012b; King et al. 1994; Rohlfing 2012.

quality of neopositivist research is usually measured in recourse to what Kvale (1995: 20) has called the "scientific holy trinity" of validity, reliability, and generalizability (also Eckstein 1975; King et al. 1994).

In contrast, reconstructive or qualitative researchers stress that academic research itself is a social practice that, because it is embedded in the social world, is unavoidably influenced by and influences its social surroundings (Denzin and Lincoln 2005a). As a consequence, qualitative researchers usually share a skepticism toward the hallmarks of neopositivist research, including the subject/object and fact/value distinctions, the assumption that theories can (or should) be tested against an independent and observable "reality," the aim of generalization beyond a specific social context, and conceptualizing social phenomena as causally related variables (Denzin and Lincoln 2003a, 2003b, 2005b).[20] As a consequence, qualitative researchers often stress the socially constructed nature of research itself and the need for transparency and, above all, for researching subjects to reflect on their own "positionality" (Cousin 2010). I return to this need below.

In terms of research design, qualitative researchers stress that it is important not to approach a substantive phenomenon under investigation with prefixed theoretical concepts, methodology, methods-based quality criteria, and so on but to develop tools adequate for the specific research problem at hand.[21] Arguably the most important consequence of a reconstructive approach is to develop a research design that is appropriate to the phenomenon under investigation instead of predetermined by universal external standards of what "good" research is (Franke and Roos 2013a; Herborth 2011; Salter 2013b). As a consequence, a good qualitative research design— including the research question, methods, and so forth—must, above all, be flexible and adaptable to the problem at hand (see Schwartz-Shea and Yanow 2012).

The study reported here was subject to a number of modifications dur-

---

20. In that sense, qualitative or reconstructive research differs significantly from neopositivist, small-n research in political science that is, somewhat misleadingly, referred to as "qualitative" (e.g., Brady and Collier 2004; Collier and Mahoney 1996; George and Bennett 2005; Gerring 2007; King et al. 1994; Mahoney 2010) but that, aside from focusing on a small number of cases, does not share the basic outlook of reconstructive research as outlined above and differs fundamentally from it in regard not only to metatheoretical assumptions but also to the criteria that distinguish "good" science from "bad" (Kvale 1995; Seale 1999; Tracy 2010).

21. Along similar lines, see Beiner 2009 (in the German language), for a highly insightful and accessible discussion of humanities research.

ing the analysis. For example, to take into account events that happened after the analysis began, the text corpus, originally including only the 11th–16th legislative periods (until 2009), was expanded to include the 17th period. The expanded corpus took into account the 2009 Kunduz air strike that, called in by a German officer, killed up to 142 people, including a considerable number of civilians, and significantly influenced the debate about the Afghanistan mission and the use of military force more generally (Noetzel 2011).

The methods employed in this study also underwent modifications. The original study was planned to use a mixed-methods design (see Creswell 2003), combining quantitative lexicometric or corpus linguistic methods with an in-depth interpretive analysis of parliamentary debates (see Glasze 2007b; Nabers 2015). Due to a number of practical and methodological considerations, the lexicometric analysis had to be abandoned in favor of an interpretive discourse analysis, for two primary reasons, the first of which was practical.[22] Lexicometric analyses require machine-readable documents, and while German parliamentary debates are available as PDFs online via the Bundestag's online documentation and information system,[23] older documents were of such poor quality that they could not be converted reliably into a machine-readable format.

Growing methodological concerns also influenced the decision to refrain from further pursuing the lexicometric analysis. Although, in principle, corpus linguistic methods are primarily descriptive in nature, gaining meaningful results (about the frequency of certain words and so on) requires a number of interpretive interventions on the part of the researcher, including, for example, excluding words that are frequently used but do not say anything about the specific character of a given document (e.g., salutations at the beginning of speeches). Such interpretive interventions are not reflected in the results of a lexicometric analysis, which consist of seemingly objective frequency counts. This absence poses significant problems in terms of reflexivity and transparency as quality criteria for qualitative research. Put simply, instead of making transparent and problematizing the researcher's "presence" (Schwartz-Shea and Yanow 2012: 95), the presentation of frequency counts deletes it, which is why I decided to abandon this part of the analysis. Such adaptations of a study's research design not only are accept-

---

22. On lexicometrics, see, e.g., Wiedemann 2013.
23. http://suche.bundestag.de/plenarprotokolle/search.form

able in reconstructive research but are commonly seen as a necessary pre-condition to produce analyses that do justice to the phenomenon under investigation.

Explanation as Articulation

What does it mean to provide a discourse theoretical explanation? Discourse theory's metatheoretical commitments affect its understanding of explanation, which differs significantly from a deductive-nomological model of explanation or even from causal explanation more generally (Diez 2001: 12–13; Hollis and Smith 2004; Sanders 2002). To begin with, in contrast to other discursive approaches that claim that discourse has a causal effect on policy action (Banta 2013; Guzzini 2011; Schmidt and Radaelli 2004), a poststructuralist approach must reject any causal connection between discourse, on one hand, and policy action, on the other. The reason is relatively straightforward: because policy action itself is meaningful, it has to be conceived of as inherent to discourse. As a consequence, claiming a causal connection between the two would be tautological (for a detailed discussion, see Hansen 2006: pt. 1).

Differences between a discourse theoretical understanding of explanation and more conventional ones also occur on a more fundamental level. Discourse theory conceives of theory-guided explanation as a form of articulation. In this context, Howarth (2005) speaks about a "method of articulation." An explanatory articulation reads a substantive phenomenon through a specific theoretical lens and, as a result, produces a specific interpretation (articulation) of that phenomenon. Like any other form of articulation, it rearranges specific discursive elements that are part of the phenomenon under investigation and establishes new relations between these elements and elements of the theory. As a result, the meaning of both "theoretical" and "empirical" elements is modified by being rearticulated in a common discursive formation.[24] A discourse theoretical explanation, then, "involves *a mutual modification of the logics and concepts articulated together in the process of explaining each particular instance of research*" (Howarth 2005: 327, italics in original).

What this means in practice can be illustrated using the example of the

---

24. The scare quotes highlight that "facts" are always "theory-laden" (Yanow 2006: 13). As such, theory and empirical "data" are difficult to neatly separate in practice.

concept of the empty signifier in this study. As noted in chapter 1, previous studies have primarily highlighted cases in which a single signifier assumed the representation of a chain of equivalent demands (e.g., Herschinger 2011; Nabers 2015; Nonhoff 2006), so that assumption served as a preliminary theoretical starting point guiding the present study. However, over the course of the empirical analysis, it became apparent that the discourse of networked security was constructed around not a single signifier but a cluster of closely related signifiers that referred to, broadly, the same demand—a blurred empty signifier. As a consequence, the theoretical framework had to be modified (drawing, at least partly, on arguments from Reyes 2005), that is, rearticulated. This example demonstrates how an actual discourse analysis leaves one with both a new interpretation of the substantive case under investigation and a rearticulated version of the theory.

Politics and Reflexivity

This very specific notion of explanation has significant consequences for the epistemic status of the research product as well as for the way in which research should be conducted, which brings us back to discourse analysis as reconstructive research. To begin with, a discourse theoretical approach entails the acknowledgment that any analysis will unavoidably produce a specific and partial representation of "reality," depending on the different theoretical, normative and social discourses in which the researching subject is embedded (Howarth 2005: 322). This limitation has a number of consequences. To begin with, it points to the necessity of epistemic modesty and skepticism. Thus, instead of producing objective truth, a discourse theoretical interpretation is formulated from a specific standpoint in time and space, presenting a local form of truth rather than a general one (Kvale 1995: 21). As Howarth points out, "discourse-theoretical interpretations can only count as 'candidates for truth or falsity,' that is, can be regarded as potentially true, if they first accord with the social ontologies and 'regimes of truth' within which they are generated" (2005: 328).

More specifically, a discourse theoretical approach requires the explicit acknowledgment of the situatedness and political nature of its own representations, very much in line with critical IR (Laclau 2014c: 134; Zehfuss 2013). On one hand, discourse analysis sets out to question assumptions that are taken for granted, which is why Nonhoff stresses that discourse analysis should be seen as an explicitly political, "interventionist form of academic work" (Nonhoff 2017b: 2; also Guillaume 2013: 29–30). On the other hand, the results produced by that analysis are equally situated, partial, and selective, highlighting certain aspects and playing down others; they are, in that sense, political and never neutral or objective (see Neumann and Neumann 2017). Like any other practice, research has to be understood as "performative" (Butler 1993: 13; Denzin and Lincoln 2003a, 2003b).[25] Indeed, as is nowhere more visible than with respect to colonialism, research itself is often implicated in the reproduction of structures of domination (Denzin et al. 2008; Hendershot and Mutimer 2018; Said 1979; Smith 2013).

Such a form of research has to implement different quality criteria than studies that seek value-neutral, objective knowledge (e.g., Geddes 2003; Goertz and Mahoney 2012b; King et al. 1994; Mahoney 2010). Here, the notion of objectivity is replaced with the notion of trustworthiness (Cousin 2010; Denzin 2009; Denzin and Lincoln 2003a, 2003b; Flick 2006: ch. 2; Seale 1999; Tracy 2010), for which transparency and reflexivity particularly play a crucial role. Thus, rather than striving to minimize the impact of the researching subject on the research process, qualitative, reconstructive, or interpretive researchers would doubt that the process could ever be sufficiently quarantined. From such a perspective, the researching subject is necessary for research to function. As Glynos and Howarth argue, "the process of articulating different elements together in order to construct a critical explanation always requires practices of judgment enacted by a particular researching subject" (2007: 183; see Bull 1966).

The many decisions involved in such research is visible even in very mundane, practical matters. One example is translation. A study that relies on German-language documents but is written in English requires that quotes are translated. In this book, all translations, if not otherwise indicated, have been done by the author, which, in itself, involves a significant amount of interpretation. Literal translations, particularly on a word-by-word basis (Wierzbicka 2013a, 2013b), often miss the exact meaning of a certain phrase or passage. Most words simply do not perfectly line up in meaning across languages. For example, the German word *Gewalt* is best translated into the English word *violence*, both referring, at least in everyday discourse (but see Galtung and Hoivik 1971; Vázquez 2011), to the infliction of direct physical damage to people (Tilly 2003: 3). At the same time, *Gewalt* can, in

---

25. This performative aspect has already been pointed out by, for instance, George, in his argument about the "world-making nature of theory" and about its character as "everyday political practice" (1994: 3).

certain contexts, refer to what in English is called *force, power, control,* or *authority*, to name but a few examples. Thus, translation often involves the decision between either a literal translation that misses the actual contextual meaning of a certain passage or a translation that departs from a faithful translation but approximates a passage's contextual meaning (Wierzbicka 2013b: 2). I have opted for translation that preserves contextual meaning, while providing original German terms in cases of ambiguity. Nevertheless, it is worth noting that the limits to "translatability" (Wierzbicka 2013b) already pose a significant challenge to objectivity (presuming objectivity were possible).[26] This challenge is but one example of how interpretation and judgment are unavoidable in research.

As a consequence of what they see as an unavoidably less-than-objective nature of research, qualitative, reconstructive, or interpretive studies place emphasis on reflexivity and transparency.[27] Reflexivity here refers to the process of critically questioning the researcher's "own sense-making and the particular circumstances that might have affected it" (Schwartz-Shea and Yanow 2012: 100).[28] At the same time, if objectivity is indeed an "illusion," as Ashley (1981: 207) claims, the consequence has to be not only to question one's own choices but to lay bare the researching subject's involvement. Thus, interpretive research requires an attitude of "criticality" (Guillaume 2013) in which the researching subject constantly questions his or her choices and recognizes their political quality. A reconstructive research logic, as it is understood here, involves a form of "auditing" as a "methodologically self-critical account of how the research was done" (Seale 1999: 468), to transparently communicate choices made during the research process and to reflect on potential biases (Salter 2013b: 15; Weller 2005).[29] Above

---

26. Whether and to what extent worldviews depend on language and, as a result, might not be understandable for speakers of other languages continues to be debated (Chakkarath 2015: 5).

27. Many (neo)positivist scholars also recognize that reality does not reveal itself directly to the researcher's eye. The main difference between researchers is in which consequences are to be drawn from this fact. Where neopositivist researchers opt for technical solutions (i.e., methods) to minimize the impact of values, preconceptions, bias, and so on, my argument would be not only that keeping the researching subject out of the research is impossible but also that the instruments employed are themselves productive rather than neutral (Aradau and Huysmans 2014; Aradau et al. 2014; Law 2004, 2009; Smith 2013).

28. Reflexivity has also received significant attention in IR (Davies et al. 2004; Eagleton-Pierce 2011; Hamati-Ataya 2011, 2013, 2018; Hoffman 1987: 232).

29. Researchers emphasizing the discursively constructed nature of the world would argue that reflexivity is a necessary ingredient for any methodologically careful

all, this auditing involves a consideration of the researcher's own positionality (Cousin 2010), that is, the different academic (e.g., theoretical, methodological, empirical) and everyday discourses (political, personal) in which
the researching subject is embedded, which will influence how a specific
researching subject interprets the research material. Prior "knowledge" (theories, preconceptions about empirical phenomena, etc.) and (often implicit)
normative commitments influence which questions we ask, which aspects
of the research material we deem important or negligible, and how we interpret the material (Herborth 2010: 278–79; Nørgaard 2008: 4).

In the following discussion, I will try, insofar as the limited scope of this
chapter allows, to alert the reader to various decisions made during the
research process, thus deliberately disrupting what might otherwise have
been a smooth scientific narrative about a coherent object (see Law 2004).
This requires a shift in gear when it comes to style, from the neutral language
of "science" to the explicit thematization of the personal experiences of the
researcher—that is, explicitly "writing the researcher(s) into their work"
(Higate and Cameron 2006: 220). It also means to abandon, to some extent
at least, "the academic voice that seems mandatory if one is to gain scholarly
legitimacy" (Doty 2004: 378).[30] Such a shift in both focus and style is, despite
increasing autoethnographic work in IR, still unusual.[31] However, it is, in my
opinion, unavoidable if the aim is to allow for a qualified assessment of the
epistemic status of the knowledge produced as partial, political, and contingent on the exercise of judgment by a specific researching subject.[32]

---

research endeavors, including quantitative research (Mays and Pope 1995: 109). After
all, God did not hand down the Correlates of War data set together with the Ten Commandments, and the current "replication crisis" is only one example of how quantitative studies are also much more subject to interpretation than is commonly assumed
(Ishiyama 2014; Pashler and Wagenmakers 2012; Schooler 2014; Silberzahn and Uhlmann 2015).

30. Law (2004: 2), for instance, discusses in detail the norm to present research
findings in a precise, concise, and consistent manner that often cannot account for
the "messy" nature of reality.

31. Over the past years, a number of special journal issues and books have focused
explicitly on autoethnography (see Bleiker and Brigg 2010; Brigg and Bleiker 2010;
Dauphinee 2010, 2013a, 2013b; Doty 2010; Edkins 2013; Inayatullah 2013b; Muppidi 2013; Neumann 2010).

32. The need to reflect on the researcher's assumptions and biases was demonstrated in no more immediate fashion than in the case of Bronislaw Malinowski, who
became rather famous among anthropologists due to his methodological contributions that stressed the importance of analyzing different cultures "from the native's
point of view" (Geertz 1975: 47) and in a general frame of "tolerance, sympathy and
empathy" (Stocking 1968: 189). The credibility of his contributions was put in serious

That said, there are limits to the extent to which such an audit can succeed. Thus, if we take seriously the assumption that subjects are discursively produced, our ability to question the very discourses that constitute us as subjects will be limited. One example of this limitation is what Fleming (2018: ch. 1) calls "racial stupidity," the fact that people living in Western countries, particularly white people, often fail to recognize racism and systemic discrimination for what it is, because they live in a white supremacist society in which racism and racial discrimination is often hidden and in which the continued oppression of people of color is usually denied or explained by factors other than racism. Any reflection on a researching subject's decisions is also limited in a more practical sense. If a complete reflection of one's own biases was theoretically possible, it would certainly exceed what can realistically be done as part of a study that itself is not primarily autoethnographic.[33] Strictly speaking, the empirical demonstration of how my specific positionality as the author of this particular book differs from the interpretation that a different researching subject would have produced would only be possible on the basis of a second study conducted by someone else. Thus, while reflexivity can help lay bare a number of potential biases and the overall political character of the analysis, it cannot empirically show how this particular analysis was influenced by the specific positionality of me as a particular researching subject as compared to a different researching subject. Nevertheless, even if demonstrating exactly how the author's positionality influenced the analysis remains out of reach, it is still crucial to critically reflect on different factors that influence the analysis presented here, including theoretical presuppositions, potential sources of bias, and so on, although the following discussion of such factors will have to remain incomplete for the practical reasons already specified.

Most importantly, any empirical findings will be influenced by the main theoretical framework of any study—in this case, discourse theory. This point seems obvious, but it is worth recalling that while discourse theory provides new insights on the empirical phenomenon under investigation here, a hegemony analysis delivers a partial and selective reading, stressing only specific aspects of what is an infinitely complex, ambivalent or even

---

doubt by the posthumous publication of his research diary, which revealed him as blatantly racist (Stocking 1968: 189).

33. Dauphinee's (2013a) autoethnographic study demonstrates how such a reflection can easily escalate into a book-length manuscript.

contradictory discourse. Aside from the main theoretical framework, other articulations and discourses, such as additional theoretical as well as normative frameworks (prior knowledge about what is commonly considered relevant in German foreign policy and in security studies more generally), unavoidably influence what is deemed relevant or negligible and how it is endowed with meaning during the analysis.

To begin with, the acceptance of out-of-area operations only seems puzzling if read against the background of various bodies of literature, most notably conventional constructivist research that would expect antimilitarism to constrain German policy (e.g., Berger 1998; Maull 1990), research questioning the utility and ethics of military operations (e.g., Zehfuss 2018), and peace research that criticizes a strong military bias in German policy on conflict prevention (see, e.g., Fischer 2004; Müller 2000), including my own previous research (Stengel and Weller 2008, 2010). My prior academic interests influence the way I approach data analysis. Thus, my concern that the acceptance of out-of-area operations as a social practice needs explanation has to do with the specific academic debates in which I am embedded. Against a different theoretical and/or normative background, other aspects of German security policy might seem in need of explanation. Thus, a number of authors are more concerned with what they perceive as an insufficient adaptation to an objectively changed security environment. Understood this way, the puzzle in need of explanation is not so much the expansion of out-of-area operations per se but why force transformation has not kept pace with international developments despite what those authors argue is an obvious need to enable the Bundeswehr to "meet future operational challenges" (Dyson 2019: 1). Both positions reflect (implicit or explicit) theoretical and empirical assumptions (regarding, e.g., the current security situation, the most important threats and their underlying causes, or the effectiveness of military operations) as well as normative commitments (e.g., about the desirability of military operations) that influence what we perceive as puzzles in need of explanation.

Aside from discourse theory itself, the research reported here has been influenced by critical studies on security and/or the military (for an overview, see Browning and McDonald 2013; Stavrianakis and Stern 2018) and by critical IR more generally (e.g., Edkins and Vaughan-Williams 2009; Zehfuss 2013), particularly feminist, postcolonial, critical geopolitical, and (broadly) poststructuralist analyses of foreign policy, violence, and interventions, which has contributed to an increasingly skeptical stance on my part regarding the military as an institution and its utility for foreign policy. Moreover, it has alerted me to the "dark sides" of Western states' foreign policies, which "we" in the West too often simply assume to be obviously "benign" (Hoffmann 2001), including ethical problems associated with (purportedly humanitarian) military interventions.[34] This tension fuels much of my interest in studying military operations in the first place and unavoidably influences how I approach out-of-area operations as an object of analysis.

The various discourses in which any researching subject is embedded will very likely make certain text passages or phrases seem more relevant (almost jumping at the researcher) while others go relatively unnoticed. For instance, reading feminist literature will alert one to constructions of masculinity and femininity, specific constructions of rationality and irrationality, sobriety and emotionality, activeness and passivity, strength and weakness, and so on (e.g., Hooper 2001). As a consequence, it is virtually impossible not to take notice of arguments that claim, for instance, that pacifism, although it might principally be morally desirable, rests on unrealistic assumptions about the world of international politics. Such statements, which are also prominent in the discourse on German security (see the discussion in chapter 3), are a clear expression of a traditional masculinist articulation of what it means to be rational and realistic.

Occupying a certain (set of) subject position(s) within a specific society means that any researching subject will unavoidably suffer from certain biases and blind spots (Berger and Luckmann 1967; Hindess 1973) as a result of assumptions that are taken for granted. Applied to this study, informal knowledge and experiences obtained by growing up as a white male in Germany mean that I will see the world in a certain way and will very likely be less observant of, for example, racial and gender discrimination, simply because, as a member of the dominant group in my society, I am not subjected to sexism or racism in my daily life (Fleming 2018). As a consequence, it is not unlikely that I will miss (without even being able to point out) certain aspects of the discourse under investigation, aspects that might immediately jump out at, for instance, a woman of color or someone living in a society at the receiving end of Western military interventions. To some extent, this problem can be mitigated (but not resolved) by engaging critical perspectives that can create "stranger-ness" (Yanow 2006: 19) by questioning

---

34. There are countless studies concerned with these ethical problems see, e.g., Chandler 2004; Lambach 2015; Muppidi 2012; Orford 1999; Owens 2003; Shepherd 2006; Zehfuss 2018.

assumptions that are taken for granted. For example, reading literature on critical race theory (e.g., Crenshaw 2011; Delgado and Stefancic 2017; Mills 2014) can increase awareness for racial discrimination. In a similar way, critical studies problematize articulations otherwise taken for granted—such as the articulation of a universal linear progress toward Western-style liberal democracy (Chakrabarty 2000), inherent in, for instance, understandings of development or notions of the Taliban being stuck in the "Stone Age" (Steinmeier 16/233, 8 September 2009: 26303)—or alert us to the political consequences of the detached, technical language employed in research (Thomas 2011). In that sense, what researchers bring to the table is not simply a hindrance to objective knowledge but can also be enabling knowledge in the first place (Yanow 2006: 19).

Moreover, as a researching subject, I am, to some extent, a product of my history, personal and professional, which will unavoidably influence my analysis, although this is not often acknowledged in IR. For example, I am deeply skeptical of the utility and normative desirability of military violence, which (in addition to my exposure to critical IR) is partially the result of my embeddedness in social relations as a private person. This embeddedness includes, for example, my upbringing in a leftist German family with parents who were members of the antinuclear, peace, and feminist movements, as well as my own experiences as a conscript in the army (see Crane-Seeber 2017). Growing up in a specific country, furthermore, influences a subject's normative commitments, and widespread antimilitarism certainly did not lessen my own skepticism. Reflexivity can help mitigate these influences, but it is doubtful that it will completely offset the influence of normative commitments.

At the same time, being embedded in everyday discourses can be an asset for the analysis. For instance, a number of my interpretations of German parliamentary discourse become possible only because of my background knowledge about German society and its norms, values, and culture. The meanings of a number of statements in the Bundestag are clear to me because of my being part of German society. For example, in a 1987 debate about a decision to deploy naval vessels to the Mediterranean, Green Party MPs called the operation a "panther's leap" (*Pantherspung*, Beer, 11/34, 16 October 1987: 2302). The meaning of the phrase might not be immediately clear to outside observers, but for me, the term is intimately linked to German imperialism, because I learned about the German Empire in high school.[35] Know-

---

35. The *Panthersprung* refers to German emperor Wilhelm II's decision to deploy the

ing this made it readily apparent to me why the comment was met with indignation—namely, because it compared the FRG to the German Empire. This understanding of local meanings is what ethnographers have to painstakingly acquire through, for instance, participant observation or interviews.

Finally, the practical circumstances under which we conduct and write up our research unavoidably have an impact on the number and kind of ideas we can pursue and on the bodies of literature we can incorporate into any given project. Although neither impact is usually a topic in social science research, practical circumstances do play a role, and I think it is important to lay bare their effect. Most notably, our current neoliberal academic system, which places emphasis on competition between researchers for a limited number of fixed-term positions, creates pressures to publish results quickly and, as a consequence, to make practical decisions in regard to which aspects to include.[36] For instance, discussion of white supremacy's role in German foreign policy should receive more attention theoretically than it has here (see U. A. Müller 2011; Muppidi 2018; Vucetic and Persaud 2018), and this study does not engage more thoroughly with feminism, postcolonialism, and critical race studies (e.g., Crenshaw 2011; Delgado and Stefancic 2017; Hawkesworth 2010; Nayak 2007) due, mainly, to the kind of practical considerations that, in theory, should be of minor concern in determining when a research project is complete.

The preceding examples are just a few select factors that can influence an analysis. If not for space concerns, a list of them could be continued almost infinitely. It is important to note that the aforementioned factors are quite subjective, in the sense that they are relevant to me as opposed to a different researching subject. Despite an explicit commitment to self-reflection and an attitude of "self-doubt" (Salter 2013a: 2), they unavoidably influence my research in one way or another.

Relativism and the Possibility of Critique

Final issues that need to be addressed here are to what extent and exactly how critique, broadly understood as "the assessment of the behavior . . . of

---

German gunboat *Panther* to Agadir, Morocco, to force France to surrender colonial domains to Germany (see Schöllgen 1998: 406–7).

36. For a discussion of the pressures associated with the neoliberalization of higher education, see, e.g., Münch 2014, as well as the third issue of *Scandinavian Journal of Management* 28 (2012).

others in relation to certain norms" (Lepsius 1964: 82), plays a role in a discourse theoretical analysis such as this one (for a discussion from a discourse perspective, see Nonhoff 2017b). Although the main purpose of this research is not primarily normative, the question of normativity still needs to be addressed, because normativity, whether implicit or explicit, cannot be escaped. Discourse analyses are usually considered part of critical social research, in the sense that they reveal the contingency of assumptions, thus opening up "thinking space" (George 1989). At the same time, to what extent critique requires the formulation of a positive normative vision continues to be debated in critical IR, with some scholars arguing that without articulating an emancipatory alternative, poststructuralist approaches could not be considered critical (Hynek and Chandler 2013). This argument ties in with debates in discourse theory about its apparent "normative deficit" (Critchley 2004; 2012).

This problem is further aggravated in the case of discourse theory, because theorists contest to what extent the explicit political stance in favor of "radical democracy," taken by many leading proponents of discourse theory, is consistent with post-foundationalism (e.g., Howarth 2008; Mouffe 2005b). The question here is how, from a poststructuralist perspective, one can justify preferring one set of normative commitments (e.g., radical democracy) over another, without slipping into methodological inconsistency. As noted above, a discourse theoretical perspective treats normative foundations as the contingent result of past discursive struggles. As a consequence, criticizing certain practices (e.g., war) on the basis of some normative principles alleged to be universally valid can hardly be seen as consistent with a post-foundational perspective.

In this context, Marttila and Gengnagel (2015: 62) have proposed that a discourse theoretical critique can only take the form of an "unmasking critique" that "reveals [subjects' conscious self-conceptions of the world] as being symptoms of subjectively unacknowledged supra-subjective structures."[37] Beyond merely questioning discourses that are taken for granted, such a form of critique has to turn its gaze on itself, because the critic is unavoidably affected by bias, most notably by the discourses in which the critic is embedded, as well as in the critic's choice of what to critique and based on which standards (Marttila and Gengnagel 2015: 63–64). What a

---

37. This proposal corresponds to Flügel-Martinsen's (2010) notion of an "interrogating" (*befragend*) form of critique.

researcher considers worth revealing is informed by normative principles. I addressed the issue of positionality above, but the issues of what to critique and how still need at least brief address.

To navigate the dilemma of either claiming a false objectivity and forgoing the possibility of critique or falling into normative ad hocism, I choose a pragmatic solution. Rather than arguing that practices like military operations are inherently problematic in ethical terms (which would imply adopting an essentialist position ad hoc), I argue, more modestly, that they are ethically problematic within the context of the normative framework accepted by (the majority of) the participants in the discourse under investigation. I use the discourse participants' own proclaimed norms as a yardstick. Thus, I rely on a form of immanent critique (Herzog 2016), pointing to contradictions within the German security discourse, that is, between the values commonly articulated by the discourse participants themselves, on the one hand, and German security discourse and foreign policy actions, on the other.

Although core values that are especially uncontested are often taken for granted implicitly rather than spelled out, discourse participants sometimes reference sedimented normative practices explicitly, particularly in contexts where the Self is delineated from an Other. The limited space here prohibits a detailed discussion of all sedimented practices, but consider, for instance, the following claim made by Chancellor Schröder shortly after 9/11:

> We insist . . . that the promises of the American Declaration of Independence are universally valid. There it says:
> "We hold these truths to be self-evident, that all men are created equal, that they are endowed by their Creator with certain unalienable Rights, that among these are Life, Liberty and the pursuit of Happiness." (Schröder, 14/187, 19 September 2001: 18301)[38]

This quote provides a partial illustration of the liberal democratic values—such as equality, democratic participation, human rights, self-determination, and nonviolent modes of dispute resolution—that make up the most funda-

---

38. In the debate, Schröder quoted a German translation, not the English original. The text cited here is not my translation of the German version but a direct quote of the original Declaration of Independence (CATO Institute 2002). However, the German translation Schröder quoted does not speak of "all men" being "created equal" but uses the gender-neutral term "all people" (*alle Menschen*).

mental normative sedimented practices relevant in the context under investigation (Kirste and Maull 1996; Risse-Kappen 1995: 499–500; Viehoff 2014; Ypi 2013). The empirical analysis in the present study (in particular, chapter 3) provides a more detailed discussion of how the German liberal democratic Self is constructed through the exclusion of an authoritarian and aggressive Other.

METHODS: CORPUS SELECTION AND CODING

Having discussed the methodological implications of a discourse theoretical approach, I now turn to the more practical matters of corpus selection and methods of data analysis.

Delineating the Text Corpus: Limits of the Discourse, Corpus Selection, and the Discursive Arena

A discourse analysis usually analyzes a limited number of "texts" to draw broad conclusions about the larger discourse in which these documents appear (Schwab-Trapp 2008: 173).[39] Crucial to avoiding what neopositivist social science would call "selection bias" or "sample bias" (see Collier and Mahoney 1996; Rohlfing and Starke 2013; Seawright and Gerring 2008)[40] is the careful selection of texts included in the corpus to be analyzed (see Jørgensen and Phillips 2002; Keller 2013). A first preliminary step in the process of corpus selection is to clearly delineate the limits of the discourse to be analyzed.

In principle, one can distinguish discourses along a number of dimensions (see table 4), including their spatial extension (local, regional, national, or transnational discourses), their respective discursive arenas (e.g., parliamentary, media, or academic discourses), topical focus (e.g., national security, medical, or development discourses), location and extension in time (e.g., medieval or modern discourses), and, finally, their degree of accep-

---

39. Although this study analyzes traditional texts (parliamentary protocols), a discourse analysis can focus on a broad range of "texts," such as images (Bleiker 2017; Hansen 2011; Shim 2014; Shim and Stengel 2017), audiovisual media (Heck 2017), music (Bleiker 2009; Franke and Schiltz 2013; Franklin 2005; Solomon 2012), or even architecture (Tallis 2020).

40. Both problems are not exactly identical, because the claim to having generated generalizable knowledge is much less pronounced in interpretive studies.

tance, sedimentation, or institutionalization (discursive orders being the most sedimented). In addition, one can distinguish between different levels of (nested) discourses and subdiscourses, which can be further subdivided along different dimensions. For instance, it makes sense to conceive of development and security discourses as part of larger foreign policy discourses (a thematic dimension) or to understand the parliamentary discourse as part of the larger political discourse, which includes official government discourse as well.[41]

It is important to keep in mind that these distinctions are both analytical and political in nature, for at least two reasons. First, the arrangement of different discourses and subdiscourses is the contingent result of articulations. Discourses develop over time as distinguishable entities, as well as change, merge, or even disappear over time. Consequently, the (historically contingent) boundaries of individual discourses have to be established during the analysis and will vary with context. Second, the discourse that becomes the subject of a specific study is produced by the researcher's articulatory practices. In reality, mainly due to their unavoidable unfixity, discourses overlap, and the boundaries between them shift as meaning is rearticulated in discursive practices. Reality is infinitely complicated, so any description will unavoidably have to simplify, which entails making decisions and exercising judgment. Thus, to a significant extent, distinct discourses are the product of a political intervention by the researcher, who, through articulation, sketches the boundaries of a specific discourse and consequently creates a specific, unavoidably partial and contingent representation of that discourse.

TABLE 4. Dimensions that distinguish discourses

Dimension	Example
Spatial extension	National versus transnational or international discourses
Arena	Parliamentary versus media discourses
Topic	Security versus development discourses
Time period	Cold War versus post–Cold War discourses
Degree of acceptance/sedimentation	Hegemonic versus marginalized discourses

---

41. While the parliamentary and governmental discourses overlap to a significant extent, they are still distinct. For instance, the parliamentary discourse includes statements made within, in this case, the German Bundestag, while the governmental discourse includes statements by cabinet members, independent of the arena in which statements take place (see Baumann 2006: 85–90).

The discourse under investigation in the present study is the German security discourse (for an overview, see table 5) , which is understood here as limited in a number of ways. To begin with, the focus here is exclusively on parliamentary debates in the German Bundestag, ranging from 1987 (to incorporate the end of the Cold War and German unification) until 2013 (to incorporate the Yugoslav wars, 9/11, and the 2009 Kunduz air strike, all of which were central for German military policy). Moreover, the scope of this study is limited to external security, excluding domestic security (let alone other understandings of security—e.g., social security).[42] The external security discourse is concerned with security threats (articulated as) originating outside a country's borders. As specific thematic discourses, security discourses are concerned with the current security situation, including the most important threats faced by a given actor, the best strategy to achieve and/or maintain security under the given circumstances, and which policy instruments to employ (i.e., how) to achieve that goal. In short, security discourses, as I understand them here, are concerned primarily with grand strategy. Traditionally, the main instrument of external security provision is the use of armed forces, which is why matters of the military are usually considered part of security policy. In this study, when I refer to the German security discourse, I mean the parliamentary discourse concerned with the external security of Germany, mainly after the end of the Cold War (while I generally acknowledge that there will be some overlap with other discourses).

The structure of the discourse can be further specified in terms of its different levels. The central focus of this study is the German security discourse, which is part of the wider German foreign policy discourse and is further partitioned into topically specific subdiscourses, such as the counterterror-

TABLE 5. Specifying the discourse under investigation

Spatial extension	Arena	Topic	Time period
National discourse (Germany)	Parliamentary discourse (German Bundestag)	External security	Post–Cold War

Note: This table purposefully leaves out the question of dissemination/sedimentation. Since the present study examines discursive change over time, the degree of dissemination and sedimentation varies over the time period under investigation.

---

42. The distinction between domestic and international, inside and outside, is not unproblematic but the contingent result of past discursive practices (Hellmann et al. 2016b; Walker 1993).

ism discourse or the discourse on conflict prevention.[43] These distinctions are not necessarily generalizable beyond Germany, nor will they find universal agreement. Most notably, peace researchers and activists will likely contest the subordination of the discourse on conflict prevention to the security discourse (which one can argue is itself an indicator of securitization). However, that subordination reflects how conflict prevention is articulated by discourse participants in parliament, as part of security policy (see, e.g., Schröder, 14/35, 22 April 1999: 5764). In this particular, historically contingent case, the discourse on conflict prevention in its current form emerged as a distinct subdiscourse only with the end of the Cold War, when peace operations became a much more prominent topic internationally due to the Iraqi invasion of Kuwait and the wars in Somalia and Yugoslavia. Figure 4 provides a simplified illustration of a possible foreign policy discourse.

In this context, it is important to briefly revisit the distinction between discourses and discursive orders. I understand discursive orders as discursive formations that have established themselves as the dominant, presumed way of understanding the world within a specific issue area, society, time period, and so on. If one particular project, such as the project of networked security, manages to establish itself as the dominant discourse, it becomes the new security order. That does not mean, however, that this security order becomes coextensive with the overall security discourse. The security discourse encompasses the entirety of articulations concerned with external security, including not just the hegemonic security order but equally subordinated and marginalized discourses, political demands challenging the current order, and so forth. To some extent, there will always be alternative articulations, marginalized discourses, and discursive struggles over meaning. Hegemony does not mean that marginalized voices cease to exist. Like the discursive order, they remain part of the overall thematic discourse.

---

43. I understand conflict prevention here in a very general fashion as "any structural or intercessory means to keep intrastate or interstate tensions and disputes from escalating into significant violence and use of armed force, to strengthen the capabilities of parties to possible violent conflicts for resolving their disputes peacefully, and to progressively reduce the underlying problems that produce those tensions and disputes" (Lund 2002: 117n6). This definition of prevention is broader than more narrow understandings of conflict prevention as interventions prior to the outbreak of violence and includes conflict resolution (during an armed conflict) and peace consolidation (after the termination of combat), encompassing "different entry points for intervention at the various stages of conflict cycles" (Kaldor et al. 2007: 278). The definition I use is in line with the understanding of crisis prevention employed by the German government (see Federal Government 2004).

Fig. 4. Example for nested (sub)discourses in a foreign policy discourse

To avoid the discourse analytical equivalent of selection bias, the corpus under investigation here includes the plenary protocols of all debates occurring between the 11th and 17th legislative periods (1987–2013) whose primary focus is on German security policy and/or the armed forces. The documents were selected from all tables of contents produced during the period under investigation. Limiting the corpus to parliamentary debates has the advantage of allowing for systematic comparisons over time, because it makes sense to assume that the content of statements will vary with each audience (Baumann 2006). The analysis is based on written transcripts of all verbal debates meeting the aforementioned criteria, including, for example, debates over specific deployment decisions, which, following a 1994 ruling by the Federal Constitutional Court (Bundesverfassungsgericht, BverfG), have to be decided by the Bundestag (see Brummer 2015; Meyer 2006; Wiefelspütz 2009). Also included are legislation regarding military policy (including the defense budget), government policy statements, previously agreed debates, matters of topical interest, and major interpellations on security policy, as well as questions submitted by individual members of parliament during question time (*Mündliche Anfragen, Fragestunde*).[44]

The analysis proceeded in two stages, beginning with an analysis of the discourse structure based on the entire corpus, followed by a more in-depth analysis of a subselection of documents (Diaz-Bone 2006; Jäger 2001). The first reason for this approach is pragmatic: the corpus under investigation spans more than 25 years, and an in-depth analysis of all documents would have exceeded the amount of manageable data. The second reason is simply that not all debates are equally insightful. For instance, many debates focus on rather technical issues (e.g., changes in remuneration, the distribution of garrisons around the country, or pension reform), rather than on the purpose of German security policy. The selection of which debates should receive more attention, which relatively less, and which could be ignored altogether was based on a comprehensive reading of the entire corpus.

Two aspects in particular deserve further attention: the question of the time frame and the focus on protocols of parliamentary debates in contrast to other "texts" (broadly understood as meaningful material). The selected time frame, from 1987 until 2013, can be explained by the aim to include important, widely recognized turning points in German foreign and security policy (see Baumann et al. 1999; Duffield 1999). These points include, most obviously, the end of the Cold War and unification, as well as important and heavily disputed military operations (e.g., Operation Deliberate Force, the first German participation in a post–Cold War combat operation, in 1995;[45] the controversial 1999 Kosovo operation Allied Force; and the

---

44. The different items of business are outlined in the Bundestag's Rules of Procedure (Geschäftsordnung des Deutschen Bundestages, GOBT), the ones relevant here in §§ 75, 76, 100–106, annexes 4, 5. Excluded were all documents that were not part of the verbal debate, such as speeches directly submitted to the protocol or minor interpellations (see Kepplinger 2009: 99–116).

45. Although it is commonly claimed in the literature that Operation Allied Force in 1999 marked the first German participation in a combat operation (Baumann and Hellmann 2001: 67; Wiefelspütz 2003: 146–47), former defense minister Volker Rühe (2011: v) has pointed out that German aircraft participated in air strikes carried out against Bosnian Serb targets, as part of Operation Deliberate Force. Thus, it is debatable whether the 1999 Kosovo intervention should really be considered the "significant landmark" (Hyde-Price 2001: 19) in postwar German history that some observers

operations in Afghanistan). Specifically, with the inclusion of the 17th legislative period, the debates about the 2009 Kunduz air strike—which, according to Noetzel (2011), triggered a reinterpretation of the German operation in Afghanistan—are included.

Although parliamentary debates are a common type of document used for discourse analyses,[46] so are, for example, newspaper articles. The decision for an analysis of parliamentary debates in the present study was primarily motivated by the fact that the Bundestag is the arena where discursive struggles over policy, particularly foreign policy, are most likely to take place, which makes parliamentary protocols especially suited for the analysis of discursive change. Although decision makers too draw on news reports for their information (in addition to classified intelligence: see Avey and Desch 2014: 238), studies on press-state relations have emphasized that "routine news" is usually initiated by the political system and that the media tends to reflect positions argued within the political system (Katz 2009: 200; for the German context, see Eilders and Lüter 2000; Sarcinelli and Menzel 2007: 330). Moreover, because of the possibility to ask a speaker questions or to comment in the form of a short intervention, parliamentary debates have the great advantage of allowing the researcher to examine contestation in practice, thus making assumptions that are taken for granted accessible for the researcher (Foucault 1982: 780). In addition, the Bundestag's role is particularly important with respect to out-of-area operations, due to the Bundeswehr's character as a "parliamentary army": any troop deployment has to be put to a vote in the Bundestag (Meyer 2006: 56–59; Wiefelspütz 2003: 134). As a consequence, the Bundestag is also the most important audience for any demands to participate in military operations.

Finally important to discuss are the characteristics of the German Bundestag as a specific discursive arena, understood as a "cross-linked ensemble of positionings within a discursive space" (Wrana 2014: 36). Articulations take place in a prestructured environment, in which sedimented practices regulate what can legitimately be said and by whom. In terms of identifying which persons count as "privileged storytellers" (Campbell 1993: 7) in the

---

make it out to be. Moreover, as Maull rightly points out, the reorientation of German security policy had taken place before the Kosovo intervention, not with it (Maull 2000: 58).

46. Examples for discourse analyses of parliamentary protocols include Dalgaard-Nielsen 2006, Ilie 2003, Martinson 2012, Schwab-Trapp 2008, van Dijk 2006a, and Wagner 2005.

German Bundestag,[47] we can distinguish between different factors, including (1) various degrees of formal and informal authority (i.e., deriving from official office or personal standing); (2) speaker positions within parliamentary and government structures, parliamentary groups and groupings,[48] and the corresponding parties (see von Alemann 2000: ch. 4); and (3) topic-specific versus general authority to speak (e.g., on behalf of a parliamentary group or the government).

In terms of formal authority, members of the cabinet have a privileged speaker position. In Germany as in other countries, foreign policy is commonly considered to fall within the purview of the executive (Gareis 2005: 36; Geis 2013a: 232; Oppermann and Höse 2011: 48). Table 6 presents an overview of the different government coalitions and chancellors during the time period under investigation. In particular, the federal chancellor (Niclauß 1987; Oppermann and Höse 2011: 50) and the heads of the relevant ministries, as well as their ministers of state and/or parliamentary state secretaries, are privileged speakers (Rudzio 2003: 300, 312; Sontheimer and Bleek 2002: 320, 322; Thränhardt 2000: 62). These include, above all, the head of the federal chancellery, the foreign minister, and the minister of defense, which govern the main ministries involved in the formulation of foreign and security policy (Gareis 2005: 38; Korte 2007; Siwert-Probst 1993: 13). However, foreign policy is increasingly an interministerial affair (Baumann and Stengel 2014; Stengel and Weller 2010; Weller 2007), and representatives of other ministries too, most notably the Federal Ministry for Economic Cooperation and Development (BMZ) and the Federal Ministry of Finance, are regularly involved in debates on security policy.

Aside from cabinet members, specific MPs also enjoy a privileged speaker position. In a primarily work parliament (see Ismayr 2013; Patzelt 1997), both the Bundestag's committees and subcommittees and the parliamentary groups' thematic working groups play an important role.[49] Committees usu-

---

47. These privileged discourse participants are similar to what are called "idea entrepreneurs" (Sjöstedt 2013: 144) or "norm entrepreneurs" (Finnemore and Sikkink 1998: 893) in constructivist studies or "[i]deology [b]rokers" in discourse linguistics (Spitzmüller and Warnke 2011: 86).

48. Following the Bundestag's Rules of Procedure (Bundestag 2014), I herein use the term *group* to refer to organized party factions within the Bundestag that have taken the 5 percent hurdle, and I use the term *grouping* for MPs whose party has not taken the hurdle but who are members of the Bundestag due to, most notably, excess mandates and have formed an association (see Rule 10 (4) GOBT).

49. The function of individual MPs was determined using *Kürschners Volkshandbuch Deutscher Bundestag* (Holzapfel 1987, 1992, 1996, 1999, 2003, 2009, 2012), a hand-

ally make decision recommendations prior to a plenary vote, and given increased specialization, those recommendations often have the character of decisions. Equally, the importance of parliamentary groups can hardly be overstated. To what extent an individual MP can participate in the policy-making process depends largely on the hierarchies within the parliamentary group, which decide, among other things, the staffing of committees and the speakers during a specific debate. Moreover, their working groups are crucial for opinion formation. As a consequence, the authority of issue "experts" and of holders of executive positions in parliamentary groups or committees (e.g., committee/subcommittee chairs, rapporteurs for specific bills, the chairs of parliamentary groups or thematic working groups, members of the executive committee, and speakers for certain issues) can hardly be overstated (Ismayr 2013).

Informal authority derives in particular from whether a particular individual or the political party with which he or she is affiliated enjoys credibility within the Bundestag. The personal standing of individual speakers is difficult to discern reliably, but the relative standing of parties during a certain period of time, which often depends on how long they have been in parliament, is often reflected in reactions by members of other parties. The Green Party (Bündnis 90/Die Grünen, henceforth Greens) entered the

TABLE 6. German coalition governments, 1987–2013

Legislative period	Time period	Chancellor	Coalition	Opposition
11	1987–91	Helmut Kohl, CDU (Kohl III)	CDU/CSU, FDP	SPD, Green Party, since October 1990 also PDS
12	1991–94	Helmut Kohl, CDU (Kohl IV)	CDU/CSU, FDP	SPD, Green Party, PDS
13	1994–98	Helmut Kohl, CDU (Kohl V)	CDU/CSU, FDP	SPD, Green Party, PDS
14	1998–2002	Gerhard Schröder, SPD (Schröder I)	SPD, Green Party	CDU/CSU, FDP, PDS
15	2002–5	Gerhard Schröder, SPD (Schröder II)	SPD, Green Party	CDU/CSU, FDP, PDS
16	2005–9	Angela Merkel, CDU (Merkel I)	CDU/CSU, SPD	Green Party, FDP, PDS/ Linke
17	2009–13	Angela Merkel, CDU (Merkel II)	CDU/CSU, FDP	SPD, Green Party, Linke

Sources: Based on Rudzio 2003: 237 and Marschall 2011: 158.

book that includes curricula vitae of German MPs as well as statistics regarding the Bundestag.

Bundestag for the first time in 1983, and the Party of Democratic Socialism (Partei des Demokratischen Sozialismus, PDS) joined the Bundestag after unification. Before 1983, the Bundestag had been marked, for two decades, by a "two and a half party system" (Decker 2013b: 27), consisting of (1) the conservatives, made up of the Christian Democratic Union of Germany (Christlich Demokratische Union Deutschlands, CDU) and its sister party, the Christian Social Union in Bavaria (Christlich Soziale Union in Bayern, CSU),[50] (2) the center-left Social Democratic Party of Germany (Sozialdemokratische Partei Deutschlands, SPD), and (3) the much smaller, liberal Free Democratic Party (Freie Demokratische Partei, FDP). The system changed with the election of the Green Party, as well as with the addition of the PDS after unification. The latter became a parliamentary group in the 14th legislative period and has since come to be known as Die Linke (Decker 2013a; Ismayr 2013; Lehmbruch 1998; Niedermayer 2013). Table 7 provides an overview of the changing names of the PDS throughout the years.

The two newcomers have had a difficult start and a long journey toward establishment, which, in the case of Die Linke, is still ongoing. In part, this difficulty has to do with the new parties' challenge to established discursive practices. During the 1980s and the early 1990s, the Greens, which emerged as a new party from the environmentalist movement and advocated pacifist positions in security policy, challenged established interpretations of the Self (e.g., questioning as imperialism Western benevolent intentions in such conflicts as the Iran-Iraq War) and demanded the abolition of NATO and the Bundeswehr (see, in particular, chapter 3), which made them inherently dubious in the eyes of more established parties. This suspicion is reflected, for example, in Chancellor Helmut Kohl's (CDU/CSU) statement in October 1995 that antiwar protests, in which members of the Greens had participated, were directed not against war but "against the freedom of the Federal

TABLE 7. Name changes of the PDS

Time period	1989–2005	2005–7	2007–
Party name	PDS (briefly Linke Liste / PDS in West Germany)	Linkspartei. PDS	Die Linke

Sources: Decker 2013b: 29; Decker and Hartleb 2007: 448; Meuche-Mäker 2005: 14–17; Moreau 2013: 147–48; Neu 2013: 317–18, 20–21.

---

50. Since 1949, the CDU and CSU have formed a joint parliamentary group (Ismayr 2013: 86–87).

Republic of Germany. . . . People of your ilk [the Greens] have made no contribution to freedom, neither today nor yesterday, . . . and they will surely
neither do so tomorrow" (Kohl, 13/65, 27 October 1995: 5566). As noted
above, Die Linke's difficulties in gaining acceptance by the established parties is due partly to the party's past. The diminished standing of Die Linke
still and of the Greens up until they assumed government power as part of
the SPD/Green coalition government in 1998 is clearly visible in interaction
between representatives of established parties and the "newcomers" (see
chapter 5).

Coding for Difference, Equivalence, Antagonism, and Representation

In line with a reconstructive research logic, the analysis for this study followed an open coding procedure, building categories from the empirical
material. At the same time, the analysis was primarily guided by discourse
theory, which brings up the question of how to relate the core concepts of
discourse theory to the research material, that is, how to "find" articulations
of difference, equivalence, antagonism, and representation in the material.
On a very general level, it is important to keep in mind that such an interpretation involves, to a large extent, intuition and judgment based on a deep
familiarity with the research material, in line with a reconstructive approach.
Often, articulations of equivalence that reach across different documents
cannot be demonstrated directly in a single document but only emerge
because of a deep familiarization with the overall text corpus. Keeping this
general condition in mind, Nonhoff's approach to hegemony analysis (Nonhoff 2006, 2008; 2019; in IR also, Herschinger 2011) offers some highly useful practical guidance for how difference, equivalence, antagonism, and representation can be translated into empirical research. Table 8 provides a
concise overview of the central analytical terms used in this book, including
both the theoretical concepts themselves and the terms necessary to translate them into categories suitable for discourse analysis.[51]

As noted above, any hegemonic process begins with the (to some extent,
chronic) dislocation of a dominant discursive order. Dislocation is the disruption of any discursive order by events that itself cannot be represented
within that order. Dislocation is a "pure event" that only becomes intelligi-

---

51. This specification is what neopositivist research refers to as "operationalization" or "measurement" (see Goertz and Mahoney 2012a). To avoid the ontological
and epistemological baggage associated with those terms, I refrain from using them.

ble if it is given meaning by becoming articulated in discourse, thus becoming a historical or discursive event (Lundborg 2012: 1). Thus, dislocation is only observable by proxy, in discourse participants' articulation, in hindsight, of some event as a turning point (Abbott 1997), crisis (Nabers 2017, 2019), or focal point (Kingdon 1995; Lanzara 1998).

As opposed to dislocation, equivalence, difference, antagonism, and representation are relations that can be empirically shown in discourse. Equivalence refers to the articulation of two demands as going hand in hand with respect to a third, excluded element. When coding for equivalence, one generally looks for statements that claim that different demands or subjects are or should be compatible, mutually dependent, or somehow united (Nonhoff 2006: 265).[52] Examples include the claims that military and civilian instruments should be combined in conflict prevention or that strengthening European security and defense policy means not to undermine NATO but to strengthen its "European pillar" (Kohl, 12/53, 6 November 1991: 4366). Equivalence can also take the form of a simple enumeration of different demands or subjects, where their compatibility is simply presumed. Equivalence is a matter of degree, ranging from weak forms, in which different demands are simply articulated as compatible (demand A does not hamper, undermine, or prevent the realization of demand B), and strong forms, in which demands are articulated as mutually dependent (demand A can only be realized in conjunction with demand B or not at all).

One example to illustrate the difference between weak and strong forms of equivalence is the articulation of Operation Allied Force (the 1999 Kosovo intervention) as compatible with international law. For instance, in April 1999, Chancellor Schröder stated "that the international legal basis for the NATO operation to contain a humanitarian catastrophe exists and that it is sufficient" (Schröder, 14/35, 22 April 1999: 2765). Here, Schröder claims not that upholding international law requires an intervention—as his predecessor Helmut Kohl claimed with respect to Operation Desert Storm, the mission to drive Iraqi troops out of Kuwait (Kohl, 12/5, 30 January 1991: 68)—but only that the mission does not violate international law. The former claim would be a strong articulation of equivalence; the latter is a weak form. Both

---

52. In this context, Laclau has drawn on the linguistic terms of syntagmatic relations of combination and paradigmatic relations of substitution to illustrate equivalence and difference, respectively (Laclau 1997: 309; 2014a; see also Nabers 2015: ch. 6). I leave this discussion aside here primarily to avoid creating confusion by introducing an additional set of concepts.

TABLE 8. Central theoretical concepts and terms

Concept	Description
Antagonism	Articulation of an Other as a radical threat to the Self that blocks the achievement of a full identity
Articulation	(a) Process of linking different discursive elements, thus modifying their identity (b) Contingent and temporary result of that practice
Blurred empty signifier	Group of signifiers that refer to broadly the same demand and together fulfill the function of representing the absent fullness of society
Contrariety	Relationship in which two elements (or demands) are articulated as contrary with respect to a third element
Demand, social	Request or claim, the basic unit of discursive structures (a) Epistemic: truth claim, implying the demand to accept the particular truth claim (something is) (b) Normative: request to establish a currently not realized desired state (something ought to be)
Discourse	(a) General: totality of all discourses in the sense of the discursive (b) Specific: specific formation consisting of different moments
Dislocation	Disrupted nature of any discursive order due to the impossibility of complete closure
Element	Signifier prior to any specific discursive articulation
Empty signifier	Demand that empties itself from its content to such a degree that it can assume the representation of a whole discourse (and, symbolically, the absent fullness of society); also often referred to as "master signifier," less often "nodal point"
Equivalence	Relationship between different moments of a discourse that articulates them, while still being different, as going hand in hand because of their common opposition to an antagonistic Other
Field of discursivity	The overflow of meaning that undermines any temporary fixation
Floating signifier	Moment that has become (partially) detached from its specific meaning within a given discourse, making it open to rearticulation
Foreign policy identity	Subset of national identity; discursively constructed, contingent, temporal, and context-dependent self-understanding about what it means to be, for example, German and what is appropriate for Germany in the context of foreign policy
Formation, discursive	Relatively stable set of articulations
Hegemonization	Process of a specific discourse establishing itself as the dominant discursive order

TABLE 8.—*Continued*

Concept	Description
Hegemony	(a) Establishment of a particularity as a universal (b) Temporary result of the process of hegemony—namely, discursive dominance
Identity	Specific content or meaning that a particular signifier (or overall discourse) assumes in a specific context
Moment	Discursive element that has been temporarily fixed within a specific discourse
Order, discursive	Established, dominant discourse within a specific thematic field, time and place
Other, radical/antagonistic	(Set of) discursive element(s) that is excluded in the context of social antagonism
Project, hegemonic	Attempt to establish a specific discursive formation built around a broad, comprehensive demand as the dominant order
Representation	Relation in which a particular assumes the role of a universal (empty signifier)
Super-differential boundary	Frontier between elements that thematically belong into a particular discourse and those that do not

versions can contribute to the formation of a successful hegemonic project, either by unifying demands (strong equivalence) or by undermining criticism (weak equivalence). In contrast, the logic of difference would stress the particular content of different demands, pointing out that they do not necessarily go hand in hand or are even contradictory. For instance, one could argue that Desert Storm had nothing to do with international law but was motivated by something else entirely (e.g., imperialism).

Antagonism is more complicated to empirically demonstrate, because it manifests itself not at the level of individual articulations but only on the level of the overall discourse (see Nonhoff 2017a). The simple reason is that the expression of (ideal-typically) two opposed antagonistic chains unavoidably extends what can be said in a single statement or even text. Helpful here is Nonhoff's notion of "contrariety" (*Kontrarität*), which refers to the articulation of two demands as opposed vis-à-vis a third element (Nonhoff 2017a: 93, 2019: 72). Consider the example of the construction of Serbian leader Slobodan Milošević as an obstacle to peace during debates about the Yugoslav wars. Foreign Minister Klaus Kinkel (FDP), for example, claimed that the federal government's aim was "to enforce a peaceful solution to the Kosovo conflict. President Milosevic did not react to the community of states' months-long efforts for a political solution. He is the person mainly responsible for the tragedy in Kosovo" (Kinkel, 13/248, 16 October 1998). Here, we can clearly see that Milosevic is articulated as contrary to the Self vis-à-vis a third element, the demand for peace. At the same time, the quote provides a glimpse at a part of what might become an emerging antagonistic frontier. Milošević is constructed as contrary not only to the federal government or the Bundestag but also to "the community of states." As a result, the federal government (for which Kinkel speaks), the Bundestag (to which he speaks), and the international community are articulated as equivalent subjects—by virtue of their joint demand for peace that is blocked by Milošević. This example is, of course, only one articulation, and it would be a stretch to draw the conclusion that an antagonistic frontier had materialized. One statement by one subject does not make an antagonistic frontier; if it did, virtually anything would constitute an antagonistic frontier, and the term would lose any explanatory value. If, however, similar articulations by different subjects accumulate across time and space and if an increasing number of demands are incorporated as equivalent on the different sides of the divide, one is approaching the construction of an antagonistic frontier.

The final element of hegemony is representation, the symbolization of the overall chain (and the common good as such) by one particular demand, the empty signifier. The empty signifier can refer to at least one of two things. First, as Nonhoff has argued (2006; 2007b; 2012), crucial for the formation of a unified project is the articulation of one particular demand as a universal remedy that promises, if realized, to overcome the radical Other/the lack as a whole. That means that the empty signifier is the only demand in the chain of equivalent demands that is articulated as contrary to all demands that are part of the equivalential chain that makes up the radical Other (Nonhoff 2019: 81–82). Again, this relation can only be determined at the level of the entire discourse, across different texts. One example from the German security discourse is the articulation of networked (or comprehensive) security as the universal strategy to overcome all of the new threats that formed after the end of the Cold War (see chapter 5). Second, an empty signifier can also name the collective subject that a hegemonic project aims to construct. One example is the formation of "the West" during the Cold War. Through the exclusion of the communist East, nations that were previously distinct—or, in the case of Germany, even hostile—were rearticulated as unified allies (see, e.g., Behnke 2013; P. T. Jackson 2006). In this example, "the West" functions as a representative for the totality of equivalent subjects and their demands. Both forms of representation can, in principle, be relevant for the formation of a hegemonic project.

At least three aspects are important in determining whether a hegemonic project has been successful. A first indicator for an emerging hegemony is that different discourse participants (particularly across party lines) begin reproducing a specific articulation, which indicates that they are identifying with a given hegemonic project. One example of such a statement is the assertion by Green MP Angelika Beer that the Greens, once in favor of abolishing the Bundeswehr (see chapter 3), had, since 1987, "accepted the Bundeswehr as an instrument of politics" (Beer, 14/124, 12 October 2000: 11886). A second indicator of hegemonization of a project is a shift in how demands are referenced, from stress on demands to be fulfilled to emphasis on past successes—that is, goals previously reached, demands already realized (Nonhoff 2012: 270, 2019: 70). Finally, institutionalization (sedimentation) is a third indicator of hegemonization. Institutionalization can happen in any number of ways, including the fixation of a project's main claims in policy papers, their legal codification, the reform or creation of new institutions, or even material changes, like changing weapons procurement (Nonhoff 2012: 266).

CONCLUSION

In this chapter, I demonstrated how a discourse theoretical approach translates into research practice. On a very general level, I argued that a hegemony analysis entails a broadly reconstructive approach that emphasizes subject adequateness, openness, and flexibility, and I provided examples of how the present analysis followed such an approach. More specifically, I outlined how using a discourse theoretical framework can be translated into categories to be used during the coding process. The following three chapters provide a detailed analysis of the hegemonization of the discourse of networked security, showing how out-of-area operations were legitimized in the process.

CHAPTER 3

Creative Destruction

The Dislocation of the Cold War Security Order and the Rearticulation of Military Force

The Federal Republic is a peaceable nation.

MP THOMAS KOSSENDEY (11/34, 16 OCTOBER 1987: 2298)

Si vis pacem, para bellum.

LATIN PROVERB

This chapter analyzes the dislocation of the Cold War security order and the subsequent rearticulation of military force during the 1990s. Its first section analyzes German foreign policy identity during the Cold War as the contingent result of the production of an antagonistic frontier between a Self and a radical Other.[1] I argue that the FRG's foreign policy identity emerged as the result of the construction of the German past and the East as a common radical Other marked by what Bundestag president Rita Süssmuth (11/228, 4 October 1990: 18016) called "tyranny" (*Gewaltherrschaft*). On the flip side of this antagonistic frontier, a mythical German Self emerged, that of the FRG as "peaceable nation," in the apt words of MP Thomas Kossendey (11/34, 16 October 1987: 2298). In line with this myth of a new—purely peaceful—Ger-

---

1. Part of the argument made here has already been made by critical studies of German foreign policy that pointed particularly to the importance of the German past as a constitutive outside for the identity of the FRG (Behnke 2006; Feldman 2003: 254; Hellmann 1999; Martinsen 2010; Zehfuss 2002). Equally, studies concerned with "the West" have stressed the constitutive role of the communist East (e.g., Behnke 2013; Campbell 1998; Dalby 1988; Hall 1992; P. T. Jackson 2006; Klein 1990; Neumann 1999).

many, as I call it, the FRG's armed forces were equally articulated not as an instrument of war but as one of peace, designed not to wage war but to prevent it from ever occurring (reflected also in the Bundeswehr's leadership philosophy of Innere Führung, see Leonhard 2019).

This chapter's second section traces the dislocation of the Cold War security discourse at the end of the Cold War, with détente and the subsequent dissolution of the Soviet Union. Without "the East" as its constitutive outside, the identity of "the West" was suddenly in question (similar, Behnke 2013), and the Bundeswehr had lost its main justification. At the same time, in the context of "new" conflicts like the Gulf War, external demands for Germany to become more actively involved in peacekeeping rendered previously compatible elements of German foreign policy identity (antimilitarism and multilateralism) into conflict (Zehfuss 2002). The result was a further dislocation of the old order and the transformation of numerous moments that were part of the old order into floating signifiers open to rearticulation.

The chapter's third section analyzes the rearticulation of military force in the ensuing discursive struggles. During these struggles, one particular hegemonic project managed to assert itself. The proponents of this project—which I call "the project of Germany's (new) international responsibility"—argued that after unification and with the emergence of new conflicts, Germany faced a new "international responsibility" (e.g., Stoltenberg, 12/27, 5 June 1991: 2033) that the country had to fulfill through participation in international peacekeeping operations. The project managed to rearticulate a number of floating signifiers in such a way that the (threat and) use of military force—once closely associated with tyranny and aggression—became compatible (equivalent) with the promotion of peace.

REJECTING TYRANNY AND WAR: THE MYTH OF THE "NEW GERMANY"

Like IR poststructuralists, discourse theory conceptualizes any identity (individual or collective) as differential but distinguishes social antagonism as a specific type of Self/Other relationship in which the radical Other is constructed as blocking the Self from achieving a pure, unadulterated presence, a fully constituted identity. While such a full presence is ontologically impossible, this chronic lack of any social structures plays out on the ontic level of specific discourses in the construction of antagonistic frontiers between a Self and a radical Other that is blamed for the Self's lack of a full identity. During the Cold War, the foreign policy identity of the FRG emerged as a result of the antagonistic exclusion of both Germany's own past and the communist East, thus combining temporal and geospatial dimensions (see also Behnke 2006: 415; Engelkamp and Offermann 2012: 236). The FRG's authoritarian predecessors and the East were articulated not as disparate but as a common radical Other, which, for heuristic purposes, I refer to as the "tyrannical Other."[2]

The exclusion of Germany's past played a constitutive role for the formation of the identity of the FRG. The military defeat of Nazi Germany in 1945 is commonly recognized, in both political and academic discourses, as a "zero hour" (*Stunde Null*) (Bald 2002: 204; Hacke 2003: 27; Jarausch 2010: 17), a crisis moment that made it necessary for Germany to radically break with its past, to reinvent itself and begin anew.[3] Thus, the FRG's identity was constructed around its radical discontinuity with the past: Nazi Germany and the German Empire were authoritarian, oppressive, externally aggressive, internationally isolated (despite any wartime alliances of the "old" Germany) and somehow deviant (the German "special path," or *Sonderweg*, being the central reference here: see Berger 2002; Jarausch 1995; Kocka 1988). In contrast, the FRG was articulated as liberal democratic, inherently peaceful and peace-loving, and fully integrated into both the West and the international community more generally. For example, in his first government policy statement in 1949, Chancellor Konrad Adenauer heralded the "becoming" of a "new Germany" ([1949] 2002: 35), opposed to and radically different from the "old" Germany of the Nazis and the monarchist German Empire.

This myth of a new Germany, a highly influential perspective in post-1945 German discourse, had become naturalized to such a degree by the late 1980s, when this book sets off, that explicit references to the past as the FRG's "origin" (Zehfuss 2002: 211) were commonly taken for granted as common knowledge familiar to all discourse participants. Prevailing as a consequence are references to the FRG's positive identity instead of its constitutive Other.[4]

---

2. I here draw on Süssmuth's representation of the German past and the communist East as instances of "tyranny" (11/228, 4 October 1990: 18016, cited above).

3. Among historians, the *Stunde Null* continues to be controversially discussed, though (see Giles 1997).

4. Another reason for the relative rarity of direct references to the past is that the

One example is a 1987 description of the FRG by Theo Waigel, deputy chairman of the CDU/CSU parliamentary group.

> The Federal Republic is a *peaceful, reliable, calculable,* and, as far as the economic performance is concerned, even strong partner in the democratic community of states. *I do not plead for an outdated [hyper]nationalism [Nationalismus],* but I reckon that a reformed sense of national identity cannot be refused even to the Germans. . . . The Federal Republic of Germany is a *democratic, liberal constitutional state [freiheitlicher Rechtsstaat]* that is externally *firmly anchored in the Western democratic community of states* and internally characterized by a multiparty system, a pluralist society, and a social, free-market system. *In the 40 years of its history, the Federal Republic of Germany has proven to be a stable factor in international relations* and internally exhibits an exemplary economic and social stability. (Waigel, 11/24, 10 September 1987: 1586, italics added)

Though Waigel constructs the FRG as "a peaceful, reliable, calculable, and . . . strong partner in the democratic community" (if explicitly only in economic terms), as a "democratic, liberal constitutional state" that was "firmly anchored" in the West and characterized in economic terms by a "social, free-market system," a second glance at this positive representation of the FRG reveals a number of implicit references to the past Other.[5] For example, Waigel's rejection of an "outdated nationalism" is a clear reference to the FRG's less-enlightened predecessors. Similarly, Waigel explicitly stressed that (only) in "the 40 years of [the FRG's] history" had Germany become a "stable factor in international relations," clearly implying that Germany's role had been quite different before that time. This radical break with the past is further reinforced through Waigel's consistent reference to "the Federal Republic" or "the Federal Republic of Germany" instead of simply "Germany." This terminology, highly common among German MPs (see, e.g., Dregger, 11/171, 26 October 1989: 12853), further underlines the general message that the Federal Republic had nothing to do with and must not be mistaken for the "Germany" of old.[6] In statements like this, the tem-

---

overwhelming majority of debates during the late 1980s focused on Cold War détente and the Soviet Other.

5. The articulation is positive in a double sense, both in the sense of "benevolent" and as opposed to the negative construction of identity through explicit exclusion.

6. For an excellent and detailed discussion of the ambiguous articulation of the

poral antagonism between the FRG and its predecessors continued to be reinscribed in the discourse, if rather implicitly. Such articulations reproduced the FRG's (mythical) identity as radically different from its past, as entirely democratic, peaceful, international—in short, morally good.

This exclusion of the past is much more clearly visible in parliamentary debates that are not focused primarily on security issues. For instance, in his first government policy statement following unification, Chancellor Helmut Kohl (CDU/CSU) stressed the "double vow" of the "fathers and mothers of our constitution": "Never again war! Never again dictatorship!" (11/228, 4 October 1990: 18019). Here, the articulation of a relation of contrariety between the FRG (signified by the "fathers and mothers of our constitution") and the past, characterized by war and dictatorship, is much more transparent. In that sense, the German past functioned (and continues to function) as a constitutive outside for the FRG, making possible the articulation of the latter as a new, radically different German state characterized, above all, by democracy and a "love for peace" (*Friedensliebe*), as FDP MP Carl-Ludwig Thiele aptly put it (12/27, 5 June 1991: 2023).[7]

In addition to the temporal dimension, the construction of a German Cold War identity involved a geospatial dimension. Equally motivated by a desire to avoid the mistakes of the past, most notably a fatal German special path,[8] the FRG had been articulated, since its foundation in 1949, as an integral part of "the West."[9] As a consequence, the East-West antagonism highlighted in previous studies also manifested itself in boundary-drawing practices in German discourse, differentiating the FRG and the West from the Warsaw Pact and, in particular, the FRG's "evil twin," the GDR. Like Germany's past, the communist East was similarly constructed as internally oppressive (tyrannical) and externally aggressive. In both cases, external aggression was articulated as a consequence of an authoritarian political system. For instance, during the speech quoted above, Waigel discussed the sources of the conflict with the Soviet Union and the possibility of overcoming it. Quoting a statement by the CSU leadership, Waigel claimed,

---

FRG as simultaneously identical with and different from its predecessors, see Zehfuss 2002; for analysis of (constructions of) war memories in that context, Zehfuss 2007.

7. The positive side of German identity has already been pointed out by a number of constructivist studies (see, e.g., Baumann 2011: 468).

8. Chancellor Kohl, e.g., insisted in 1990, "There won't be any German special paths or [hyper]nationalist solo actions [*nationalistische Alleingänge*] in the future either" (11/228, 4 October 1990: 18026).

9. For the production of Germany as an integral part of "Western Civilization," see P. T. Jackson 2006.

In reality it is not primarily about missile numbers on both sides but *about a political change in the Soviet Union that would permanently realize world peace.* Abandonment of world revolution, abandonment of strategic superiority, abandonment of ideological and imperialist interference in the countries of the Third and Fourth Worlds, freedom inside the Soviet area, peace, and non-aggression of its foreign policy are the decisive topics.

If the SPD was willing to [conduct] a sober analysis of the situation, it would have to come to the same results. . . . Anyone who denies his [sic] peoples [their] freedom and self-determination, who today still wages a colonial war in Afghanistan, and who threatens Europe with a triple superiority in terms of conventional weapons cannot suddenly be renamed a partner. (11/24, 10 September 1987: 1584, italics added)

Two aspects of this claim are of particular importance here. First, "political change in the Soviet Union" is proposed as the only viable solution that could bring about "world peace" (a vision if there ever was one). Because external aggression was seen as the result of authoritarianism, only democratization could lead to a lasting peace (see also, e.g., Kohl, 11/24, 4 October 1990: 1596, 1597). Second, the past and the communist East are articulated as equivalent, as a common Other. While Germany and the West had left their aggressive, imperialist policies in the past, the East continued to pursue such policies well into the present. Internal oppression (the denial of "freedom and self-determination") and external aggression ("world revolution," "colonial war") were two sides of the same coin.

These articulations gained additional credibility by drawing on established discursive patterns. The articulation of a Western Self firmly located in the present compared to a backward non-Western Other stuck in the past has been a prominent argument in Western discourses at least since colonialism (see, e.g., Bhabha 1984; Chakrabarty 2000; Chandra 2013; Darby 2009; Said 1979; Spivak 1988). Not only are similar articulations immediately appealing because of their familiarity, but they equally reinforce the assumption that "we" can legitimately make decisions on "their" behalf (in this case, prescribing Western-style democratization as the only acceptable way forward). After all, since "we" are located in front of "them" on the linear path to progress, we already know the right way forward. In addition, the preceding quote of Waigel shows the importance of gender discourse in producing credibility. Drawing on hegemonic constructions of masculinity as rational and emotionally detached (e.g., Carver 2006; Connell 2005; Hooper 2001; Tickner 1988), Waigel dismissed alternative articulations by the SPD as insufficiently "sober," as motivated by (irrational, utopian) wishful thinking instead of being based on a realist(ic), cold, hard look at reality.

Underlying both the historical and the East-West antagonism is a more fundamental opposition, between freedom and tyranny. In a somewhat simplified version of the democratic peace thesis, tyranny was articulated as the reason for war, and democratic regimes were claimed to be inherently peaceful.[10] As a consequence, the populations of both the FRG's predecessors and Warsaw Pact countries were articulated as part of the Self rather than the Other. This perspective is clearly visible in Waigel's claim that the Soviets denied their peoples "freedom and self-determination." The populations in Warsaw Pact countries were articulated as oppressed, rather than as on the side of the ruling class.

Similar arguments persisted with respect to the German population before 1945, referred to as victims of the Nazi regime rather than, for example, "willing executioners" (Goldhagen 1996).[11] Even members of the Wehrmacht were articulated not as accomplices but as a neutral instrument abused by the Nazi elite,[12] as illustrated by a debate about the highly controversial "soldier rulings." The specifics of the debate, which was aptly titled "The Peace Duty of the Bundeswehr" (11/171, 26 October 1989: 12853–69), need not concern us here. What matters is that it was triggered by the abatement of legal proceedings in the case of a physician who had publicly called all soldiers "potential murderers" and had consequently been charged with sedition and defamation (quoted in Perger 2002: 123).[13] During that debate, FDP MP Werner Hoyer claimed that the ruling was against

---

10. For more-detailed discussions on the democratic peace thesis and political practice, see Geis and Wagner 2011; Russett 2005.

11. Such articulations of victimhood, if not as explicit, can be traced back to Adenauer ([1949] 2002: 35) but continue to exert influence in German political discourse. For instance, in a debate in 1995, Foreign Minister Klaus Kinkel (FDP) claimed that the US, the UK, and other states had "liberated us from the Nazi dictatorship" (13/48, 30 June 1995: 3957). I have not provided a quote from the late 1980s here because the lessons of the past rose to prominence again only in debates about out-of-area operations (see Zehfuss 2007).

12. See also Zehfuss's (2007: 11) discussion of this "myth of the 'clean' Wehrmacht."

13. The statement is a modified quote of Kurt Tucholsky, who, in 1931, had called soldiers murderers. Already, Tucholsky's statement led to legal proceedings, because it was seen as slanderous in regard to the Reichswehr, the Weimar Republic's armed forces (Perger 2002).

those who, during the Second World War,—*abused by a certainly murderous regime*—have *fulfilled their duty to serve* in all conscience, as well as their relatives, whose feelings are being gravely hurt by this ruling. My father, who has unwillingly served for ten years of his life in the army, who has been *forced by an unlawful regime* to sacrifice his youth, for me was and [still] is neither a potential nor an actual murderer." (11/171, 26 October 1989: 12861–62, italics added)

The quote clearly demonstrates that the antagonistic frontier is located between the Nazi regime's tyrannical elites, who alone are to blame and who are firmly located in the past, and the German people, including the soldiers, who, "abused" by a "murderous regime," were victims of oppression and abuse and, as a result, could be part of the FRG and the present.

The articulation of the past and the East as instantiations of the same phenomenon is most clearly visible in an opening address given in 1990 by Bundestag president Rita Süssmuth (CDU/CSU) to commemorate the "victims of tyranny" (*Opfer der Gewaltherrschaft*—literally, "victims of rule by violence") (11/228, 4 October 1990: 18016). Süssmuth made clear that "tyranny" included not just Nazi Germany but also the Soviet Union and the GDR.

More than 100 members of the *Reichstag* [the parliament of the Weimar Republic] met their death, perished *in the jails and National Socialist concentration camps*, were murdered, executed, driven into death, died from the consequences of imprisonment. Many became the victims of *communist cleansing in the Stalin era. We commemorate all victims of the National Socialists, and we commemorate the victims of the SED regime*, the victims of [the] wall and barbed wire." (Süssmuth, 11/228, 4 October 1990: 18016, italics added)

Here, "tyranny" functions as an empty signifier representing a (antagonistic) chain of equivalence including both Germany's past and the Warsaw Pact, united in their persecution of the innocent. Süssmuth did not explicitly equate communism with National Socialism, but Nazi Germany, the Soviet Union, and the GDR were still articulated as different manifestations of tyranny as a single phenomenon and interwoven into a common Other to the liberal democratic FRG and "freedom" more generally.

The result of the construction of this aggressive and tyrannical Other was

FRG/the West

**German Empire/
Nazi Germany/
Communism**

democratic
liberal
peaceful/antimilitarist
Western/integrated
multilateral
reliable/predictable

authoritarian
oppressive
aggressive/militarist
deviant/isolated
unilateral
unreliable/unpredictable

Fig. 5. The FRG and its tyrannical Other. (Based on Nabers 2009: 206.)

the (implicit) production of a mythical Self. The tyrannical Other was articulated as internally authoritarian, oppressive, militaristic, and marked by hypernationalism, while externally aggressive, belligerent, isolated, unpredictable, and deviant vis-à-vis the common European path. In contrast, the FRG was liberal democratic, peaceful, firmly integrated within the community of Western democracies and the international community more broadly, predictable, and only moderately—indeed, reluctantly—patriotic (see fig. 5).

Moreover, as the past and the East were regarded as belligerent by virtue of their authoritarian political system, the FRG was considered peaceful because of being (liberal) democratic. Süssmuth's speech cited above is illustrative here too. Immediately following a minute of silence, in which the Bundestag commemorated the victims of tyranny, Süssmuth stated that "free parliamentarians and free parliaments are the guarantors for the protection of human rights, freedom and human dignity." She argued that Germany's own past, "our own history," imposed "a special responsibility" upon Germany: namely, "to champion peace and the protection of human rights," both at home and abroad (11/228, 4 October 1990: 18016).

In sum, the construction of an antagonistic frontier between the Self and a tyrannical Other produced a positive identity according to which the FRG was and should remain liberal democratic, observant of the rule of law, well integrated internationally, and, above all, inherently peaceful. How engrained this mythical notion of the peaceful, benevolent Self was becomes clearly visible in moments of contestation. One example is the deployment in 1987 of German naval vessels to the Mediterranean to relieve allied ships that, in turn, were to be sent to the Persian Gulf to clear sea mines deployed during the ongoing Iran-Iraq War. During the parliamentary debate, Green Party MPs decried the mission as a "particularly despicable form of gunboat policy" (Mechtersheimer, 11/34, 16 October 1987: 2297) and referred to it as a "panther's leap" (*Pantherspung*, Beer, 11/34, 16 October 1987: 2302). Both phrases, "gunboat diplomacy" and "panther's leap," have very specific historical connotations clearly related to German colonialism, and by using these terms, the Greens challenged the strict separation between a benevolent German/Western Self and its own imperialist past, as well as the claim that the mission served peace. Not surprisingly, the remarks were met with near-universal reproof by members of the government coalition. For instance, Defense Minister Manfred Wörner (CDU/CSU) and MP Klaus-Dieter Uelhoff (CSU/CSU) unanimously charged Green MPs with demonstrating "a deplorable ignorance" (Wörner, 11/34, 16 October 1987: 2304) and having "no clue" of history (Uelhoff, 11/34, 16 October 1987: 2302). Clearly, in comparing US and German policy to imperialism, a line had been crossed between, on one hand, acceptable informed, rational, and reasonable comment and, on the other, unfounded, ideologically infused, unreasonable speculation and outright blasphemy.

That such reactions are not an exception is illustrated by another example. In a debate about Desert Storm in January 1991, the chairman of the Linke Liste / PDS parliamentary grouping, Gregor Gysi, charging Chancellor Kohl with "hypocrisy" for not calling Desert Storm a war, demanded that all wars, including Western ones, be regarded as "criminal" (12/3, 17 January 1991: 52, 53). In response, Süssmuth, then Bundestag president, formally called Gysi to order, arguing that "no one in this house [the Bundestag] should" cast doubt on any other MP's "will to peace" (*Friedenswille*) (12/3, 17 January 1991: 53). Not only was Gysi's argument not given serious consideration, but he was formally reprimanded for suggesting that any action in which the West might be involved could legitimately be called war or be motivated by any goal other than peace. This clearly demonstrates how deeply ingrained the notion of a benign, peaceful Self was.

Nevertheless, its sedimentation notwithstanding, the myth of a new, peaceful Germany is just that—a myth. The construction of an antagonistic frontier is an ideological operation, because the binary representation of reality it provides is only possible by ignoring a number of heterogeneous elements that fit in neither of two camps. This applies to both the East-West opposition and the articulation of a radical discontinuity between the FRG and its past. For example, the argument that the central organizing principle of the Cold War world was the antagonism between a liberal democratic West and a communist East rests on a denial of elements like the Non-Aligned

Movement (see Miskovic et al. 2014), and only by denying such elements can a representation of "the world" as split between the West and the East be upheld.[14]

In a similar fashion, the historical antagonism rested on the ignorance of a number of elements that, if considered, would significantly complicate the story. One quite obvious example is the Weimar Republic, which was neither authoritarian nor aggressive in the sense of Nazi Germany or the German Empire but does not fit into the present either.[15] Even more interesting in the context of this study is a series of elements that complicate the story of a radical discontinuity with Nazi Germany. The most obvious element of continuity is the German population itself, which obviously was not replaced as a result of denazification. Notwithstanding claims in parliamentary discourse that ordinary Germans had been victims of the Nazi past and had been "liberated" by the allies (Kinkel, 13/48, 30 June 1995: 3957), historians continue to debate the exact degree of complicity of the German population, a majority of which at least passively endured, if not openly supported, the Nazi regime (see Longerich 2006; Schrafstetter and Steinweis 2016). Another element that puts the articulation of a radical break into doubt is the level of elite continuity between Nazi Germany and the FRG (see Edinger 1960). A number of the latter's institutions (particularly security ones) were built with the help of and later staffed with members of the old elite, including not only the Federal Intelligence Service (Crome 2007; Wegener 2007) but also the Bundeswehr, which was established by a group of former Wehrmacht generals. This continuity was made possible by the aforementioned articulation of the German military as a neutral instrument that had been abused by the Nazi elite (Bald 2002; Molt 2007; Sangar 2015). These examples show that, far from being objective or accurate, the articulation of a radical discontinuity with the past is ideological, as is the emerging myth of the FRG as a completely peaceful, just, and free society, ruled by a purely democratic state.

Despite its ideological character, this discursively produced identity regulated the realm of acceptable policies, including, in particular, which role

---

14. Klein (1990: 319) locates such elements in "that liminal space which escapes Western 'identity.'" He also lists, among others, Balkan guest workers in Germany or Native Americans in Canada.

15. Commonly said to have been a democracy without democrats (Bracher 1964; critical, Reckendrees 2015), the Weimar Republic provided a negative example for the "fortified democracy" (*wehrhafte Demokratie*) of the Basic Law (Schliesky 2014).

German armed forces could legitimately play. In fact, the set of appropriate tasks for the Bundeswehr was strictly limited, which made the maintenance of armed forces compatible with an alleged inherent German peacefulness in the first place (similarly, Clemens 1993: 235). Most notably, the use of military violence was firmly located on the other side of the antagonistic frontier. In parliamentary speeches, the threat and use of military violence was linked to the Warsaw Pact, whereas Western armed forces were articulated as purely defensive in nature. For instance, Defense Minister Wörner (CDU/CSU) claimed in 1987,

> We clearly do not want a preponderance compared to the Warsaw Pact. *What we want is only the amount of security and defense that allows us to tell our citizens: You can be safe from a war, and also tomorrow you will be able to live in freedom.* (11/41, 24 November 1987: 2778, italics added)

While Soviet armed forces were articulated as a threat to Europe and the West, the purpose of NATO forces was only to ensure the safety of "our citizens." Similarly, Wörner's successor, Gerhard Stoltenberg, pointed out in 1989 that "NATO and the Bundeswehr have since their foundation threatened no one" (11/182, 7 December 1989: 13986). The purely defensive character of the Bundeswehr is stipulated in Article 87a of the German Basic Law (Grundgesetz, GG),[16] which limits the range of tasks beyond defense to those "expressly permitted by this Basic Law," the only such task being the explicit authorization to enter a "system of mutual collective security" as specified in Article 24 GG (see Jaberg 2008: 84–87).

Moreover, like German identity in general, the identity of the Bundeswehr was constructed in explicit opposition to its authoritarian and aggressive predecessors, the German Empire's Imperial German Army (Deutsches Heer) and Navy (Kaiserliche Marine) and the National Socialist Wehrmacht. As Defense Minister Stoltenberg claimed in 1989,

> *All preceding armies were above all intended . . . to wage war.* The Bundeswehr is the first conscript army in a German democracy. It gains its political rational and moral justification from the mission to *prevent war* and to preserve peace and freedom. (11/171, 26 October 1989: 12857, italics added).

---

16. I refer here to the official English translation published by the Bundestag (2010).

With this claim, Stoltenberg explicitly articulated a relation of contrariety between the Bundeswehr, on one hand, and the Imperial German Army and the Wehrmacht, on the other. The reference point for the articulation of contrariety is the issue of war or peace, respectively. Thus, Stoltenberg pointed out that as opposed to its predecessors, whose purpose had been warfare, the Bundeswehr's prime objective, its raison d'être, was the prevention of war, to "preserve peace and freedom."[17] On a different occasion, Stoltenberg similarly claimed that the soldiers of the Bundeswehr had "to practice the case of emergency [Ernstfall] to prevent the case of emergency" (11/182, 7 December 1989: 13988).

Such articulations were not an exception but the rule. Service in the armed forces was presented as a "peace service" (Friedenssdienst) (Hoyer, 11/171, 26 October 1989: 12861), and the Bundeswehr's mission was, above all, a "peace mission" (Friedensauftrag) (Zumkley, Däubler-Gmelin, 11/171, 26 October 1989: 12862, 12864). Alfred Biehle (CDU/CSU), chairman of the defense committee, even called the Bundeswehr the "biggest citizens' initiative for peace" (11/171, 26 October 1989: 12860), and his colleague Bernd Wilz (CDU/CSU) referred to it (mocking the peace movement) as "the most convincing peace movement that we have ever seen on German soil" (11/171, 26 October 1989: 12868). In short, the Bundeswehr was articulated as an "army of peace" (Friedensarmee) (Biehle, 11/182, 7 December 1989: 14008).

This articulation could not be challenged legitimately, as the debate on the soldier rulings (mentioned above) illustrates. The court's decision to acquit the charged physician was met with severe critique by members of the established parties, and the comparison between soldiers and murderers was commonly seen as "insulting" (e.g., Däubler-Gmelin, 11/171, 26 October 1989: 12864). By questioning the relationship of equivalence between the Bundeswehr and peace (or, more generally, morality), the physician's statement had violated the limits of the discourse. After all, among the armed forces (Streitkräfte in German, which literally translates to "quarrel forces" or "dispute forces"), the Bundeswehr was articulated as a type specifically intended to never actually quarrel, and the specific articulation of the Bundeswehr as an "antiwar army" made the maintenance of armed forces credible against sedimented practices of peacefulness.

The debate about the soldier rulings equally illustrates how security discourses draw on gender discourse for legitimation, most notably on what Iris

---

17. This objective has already been pointed out in the previous literature: see Zehfuss 2007: 4.

Marion Young (2003) has called the "logic of masculinist protection." Following this logic, the figure of the soldier is constructed as a selfless, heroic figure who risks his or her life to protect the (helpless) citizen and who, in turn (at least implicitly), demands gratitude and obedience. The power of this construction, Young argues, rests on its similarity with traditional understandings of the family, in which the father protected his family and demanded obedience in return. Not surprisingly, the German security discourse is ripe with references that portray military service as a selfless act for which the public (including members of the Bundestag) should be grateful, and criticism is met with fierce resistance. For instance, in the debate about the soldier rulings in 1989, the chairman of the CDU/CSU parliamentary group, Alfred Dregger, explicitly expressed his gratitude to "our soldiers" and even compared the court's decision to acquit with the "rulings that have contributed to the demise of the Weimar Republic" (11/171, 26 October 1989: 12853). This expression is highly relevant here because the figure of the "warrior-protector" (Carver 2008: 71) stands in the way not only of a critique of the armed forces themselves, which seems ungrateful, but also of security policy more generally. In practice, a critique of (military) security policy can hardly be separated from a critique of the military. Thus, the logic of masculinist protection works to delegitimize any critique of security policy. The demands for gratitude and obedience do not come from the soldier, though, for the soldier is not authorized to speak in the Bundestag any more than ordinary citizens. Instead, politicians demand obedience and gratitude on the soldier's behalf, independent of what the soldier's actual wishes might be. In that sense, the soldier is a mute hero who is spoken for rather than heard. That situation does not make the argument any less effective, however.

The view of the army of peace as a passive deterrent became increasingly complicated during the 1990s with the Iraqi invasion of Kuwait and the outbreak of wars in Somalia and Yugoslavia. Nevertheless, as I will show in the remainder of this book, the imaginary of the "army of peace" remained highly influential. One reason out-of-area operations became possible was that their proponents managed to articulate the army on operation as still being essentially an instrument of peace.

THE DOUBLE DISLOCATION OF THE COLD WAR SECURITY ORDER

In the late 1980s and early 1990s, two sets of dislocatory events—Cold War détente and the subsequent dissolution of the Soviet Union, on one hand,

and the emergence of new conflicts, on the other—led to the emergence of political practices challenging the dominant order. Discursive struggles then ensued between different projects seeking to provide an authoritative representation of the situation and of the best response. The most successful of these projects, which can be called "the project of a new German international responsibility" (or "the project of Germany's international responsibility"), succeeded in rearticulating former elements of the old order in such a way that military operations became possible (Stengel 2019a).

Already during the late 1980s, the old order that articulated the Soviet Union as a radical Other came under increasing pressure because of glasnost and perestroika, a series of reforms implemented by the Soviet leadership of Mikhail Gorbachev from the mid-1980s onward, to open political institutions and liberalize Soviet society (see Åslund 1992; Hewett and Winston 1991; Linz and Stepan 1992). This program coincided with "new thinking" in Soviet foreign policy, leading to advances in arms control negotiations between the US and the Soviet Union (Risse-Kappen 1991, 1994), many of which were based on wide-ranging concessions made by Gorbachev (Dean 1987). These developments, which called "into question the very idea of a Soviet threat, or at least of an implacable one which can be countered only by the reciprocal threat of overwhelming counterforce" (Klein 1990: 320), destabilized the East-West antagonism on which Western identity rested. Moreover, it pulled the rug out from under the Bundeswehr.

In the late 1980s, as a result of growing dislocation, more and more MPs began to question the dominant order, particularly the need for the continued maintenance of a massive conventional deterrent. Already in the early 1980s, SPD members like Egon Bahr had begun to advocate more cooperative security arrangements with the Soviet Union, familiar under the heading of a "common security."[18] At the end of the decade, SPD MPs openly criticized what they saw as the CDU/CSU/FDP coalition government's continued efforts "to beat the other side to it in the arms race" (von Bülow, 11/182, 7 December 1989: 14005). With the end of the Cold War and the dissolution of the Soviet Union in December 1991, SPD MPs argued that the federal government should "finally draw practical consequences" from the "changes in the security political landscape" that had left "Germany's military security endangered less than ever," instead of continuing its "as-if policy," that is,

---

18. The concept was proposed by the so-called Palme Commission, of which Bahr was a member. See Independent Commission on Disarmament and Security Issues 1982, as well as the discussion in Risse-Kappen 1994.

proceeding as if the Warsaw Pact still existed (Kolbow, 12/70, 16 January 1992: 5882). They proposed that the FRG should take steps toward disarmament and a restructuration of the armed forces to achieve a "structural inability for attack" (Däubler-Gmelin, 11/171, 26 October 1989: 12864).

Critique also came from members of the Green Party, which gave voice to demands of the peace movement (Conze 2010; Probst 2013: 166–67). The challenge provided by the Greens was twofold. First, in light of Cold War détente and its later end, Green Party MPs demanded nothing short of the "disintegration of NATO, [the] reduction of the Bundeswehr [*Bundeswehrabbau*], troop withdrawal and demilitarization" (Mechtersheimer, 11/182, 7 December 1989: 13997).[19] Instead of sticking to current defense planning, they argued, the government should develop a comprehensive plan for disarmament, including conversion programs for the arms industry and economic planning for municipalities economically dependent on garrisons (Mechtersheimer, 11/182, 7 December 1989: 13999). Second, the Greens provided a fundamental critique of "realist" security policy based on such notions as the balance of power and the utility and moral desirability of armed forces.[20] In contrast to the dominant order, according to which the Bundeswehr was equivalent with peace and German security, the Greens, explicitly questioning whether the German government's (and the West's) policies really contributed to German security,[21] understood military means, on one hand, and peace and security, on the other, as contrary, not equivalent, with respect to their connection to violence.

The debate about the soldier rulings is again illustrative here. Green MPs condemned the Bundeswehr as an institution deeply implicated in the production of violence. For example, during the debate, Wilhelm Knabe, a veteran of the Second World War, argued "that armies can be deployed as murder machines" because soldiers were "trained" like dogs (*abgerichtet*) to kill. Directly challenging the equivalence between the military and secure peace,

---

19. Green MP Alfred Mechtersheimer even publicly entertained the idea of a German "redisarmament" (*Wiederentwaffnung*), a return to the pre-1955 situation, when the Bundeswehr did not exist (Mechtersheimer, 11/182, 7 December 1989: 13999).

20. The term *realist* here refers not only to IR theory but to a specific discourse that also informs political practice, what Bell (2002) calls "foreign policy ideology." While I have a different understanding of ideology, I concur with Bell's basic argument.

21. In autumn 1982, largely in reaction to NATO's Double-Track Decision that threatened the stationing of nuclear intermediate-range ballistic missiles in Germany, hundreds of thousands protested against the nuclear arms race, in the largest mass protest in the FRG's history (Brockmann 1994: 281).

Knabe claimed that in the actual event of a defensive war, the Bundeswehr
would not succeed in protecting Germany but would instead inadvertently
"destroy [our] own country" (11/171, 26 October 1989: 12856). Others
repeated the link between soldiering and murder, arguing that "politicians
who [. . .] sign the doctrine of massive nuclear retaliation" would be potential murderers (Mechtersheimer, 11/171, 26 October 1989: 12861). With unification in 1990, the Greens received additional backup from the newly
elected PDS / Linke Liste, whose MPs were equally critical of the Bundeswehr,
calling it "superfluous" (Lederer, 12/70, 16 January 1992: 5886).

The government coalition's response was mainly ideological, attempting
to gloss over the lack in the structure and to repair the dominant order
despite its increasing dislocation. For instance, in a debate in November 1987,
Defense Minister Wörner (CDU/CSU) argued that as long as the Warsaw Pact
had

> its tanks, its canons, its guns, its aircraft brought to bear against us, we cannot say:
> We live in a landscape of common security. At the moment our problem is that
> we have to maintain our security against this threat. . . . Also under [the leader-
> ship of Mikhail] Gorbachev the Soviet Union modernizes its potential without inter-
> ruption. (11/41, 24 November 1987: 2775)

Here, Wörner clearly reinscribed the East-West antagonism into the discourse, emphasizing the continued aggression of the Soviet Union, as a result of which a realistic foreign policy (as opposed to the irrational, irresponsible utopianism of common security) had to maintain a credible deterrent.

At the end of the Cold War, these arguments began to shift from the
Soviet threat toward an emphasis on a general uncertainty about the future.
This uncertainty was most clearly expressed by CDU/CSU MP Otto Hauser,
who argued, in a 1990 debate about the future development of the
Bundeswehr, that "conflicts are not avoided by forgoing any opportunity for
defense [altogether]. . . . No one would abolish the fire brigade just because
there has not been a fire in the past 40 years" (11/207, 26 April 1990: 16303).
Obviously drawing on gendered discourses for credibility, Wörner's successor as defense minister, Gerhard Stoltenberg (CDU/CSU), similarly stated
that a "responsible security policy" required "steadfastness, adherence to
one's principles, and also a certain amount of serenity with regard to short-
term fluctuations of the *zeitgeist*" (11/207, 26 April 1990: 16312, italics added).
Stoltenberg even debated with Erwin Horn, the chairman of the SPD delega-

tion in the defense committee, about whether the Warsaw Pact still existed at that time. To Horn's interjection "The Warsaw Pact is long gone," Stoltenberg responded, "No, it is not gone!" (11/207, 26 April 1990: 16312).

War outbreaks in the Gulf, Somalia, and Yugoslavia (Kusow 1994; Lockyer 2010; Prunier 1996; Weller 1992), as well as the Red Army's attack on a TV station in Lithuania (a move that many feared as the first sign of a Soviet crackdown on nationalist movements; see Mandelbaum 1992: 166n2) and continuing instability in what would soon be the former Soviet Union (Åslund 1992; Koslowski and Kratochwil 1994; Lieven 1994; Rich 1993), were enlisted as evidence for the general unpredictability of international politics and for a continued lack of peace and security in the world, as a result of which military security provision was still necessary. For instance, in 1993, the CDU/CSU parliamentary group's speaker for disarmament, Karl Lamers, argued that Germany would need NATO "also in the future," as "the events in Russia, anxiously observed by us, very clearly show" (12/150, 26 March 1993: 12867).[22] Similarly, Wolfgang Schäuble, then chairman of the CDU/CSU parliamentary group, argued in January 1992 "that also a new, global order" would not be one in which states could "relinquish . . . military means" (12/70, 16 January 1992: 5899). The Soviet Union might have vanished, the coalition argued, but in a world still characterized by uncertainty about the future, Germany had better pay its insurance premiums.

At the same time, the international community's main response to the perceived lack of peace were military interventions (e.g., Operation Desert Storm in the Gulf, the intervention in Somalia, and the series of interventions in former Yugoslavia), and Germany's allies voiced demands for troop contributions. The result was the so-called out-of-area debate, which began with Operation Desert Storm in 1991 and unfolded in the context of the UN peace operation in Somalia (UNPROFOR) and a series of interventions in the Yugoslav wars. The Yugoslav interventions, beginning with missions to monitor and later enforce a naval embargo against Serbia and Montenegro and a no-fly zone over Bosnia, culminated in Operation Allied Force, the 1999 Kosovo operation (for an overview, see Philippi 1997).

---

22. In March 1993, Russian president Boris Yeltsin and the parliament were in a mutual stalemate, with communists in the parliament and the Congress of People's Deputies effectively blocking Yeltsin's reform policies. This impasse was only resolved in September, when Yeltsin announced the dissolution of the congress and parliament as well as new parliamentary and presidential elections (which had to be enforced by the armed forces) (Easter 1997: 192–94).

Germany's decision not to commit troops to Desert Storm, due to, as Defense Minister Stoltenberg put it, the "peculiar constitutional situation and interpretation" (12/2, 14 January 1991: 38), was met with harsh criticism (Brenke 1994; Kinzer 1991; Schemo 1991), despite significant German financial contributions and a deployment of German air force units to Turkey to deter an Iraqi attack (Brenke 1994). The external demands for troop contributions increasingly problematized a relation of equivalence, self-evident until then, between antimilitarism and Germany's international integration (see also Dalgaard-Nielsen 2006; Longhurst 2004)—that is, integration in Western institutions like NATO or the EU as well as in the UN, the preference for multilateralism, and a commitment to international law. As a consequence, different elements of German foreign policy identity, itself the product of a strict antagonistic frontier between the FRG and its tyrannical past, suddenly were not equivalent under all circumstances anymore. Zehfuss observes,

> The FRG could take part in international military operations and thereby, on the one hand, confirm its integration with the West and therefore its overcoming of the Nazi past, whilst, on the other hand, making use of its military beyond defence and therefore recalling the militarist practices of the Nazi state. (2002: 209)

As a result of this dilemma, the relatively clear antagonistic frontier between the past and the present became blurred, and the Cold War order was further disrupted. Specifically, both the relations of equivalence between the different moments within the two opposing chains (say, between peace and multilateralism) and the relations of contrariety between the moments of one chain and the moments of the other (e.g., between peace on one side of the frontier and the use of force on the other) were weakened.

MAKING OUT-OF-AREA OPERATIONS POSSIBLE: THE PROJECT OF GERMANY'S INTERNATIONAL RESPONSIBILITY

While, on the basis of antimilitarism and its institutionalization in the Basic Law, the Green Party, the PDS, and, initially at least, the SPD opposed participation in out-of-area operations, the CDU/CSU and FDP argued that the FRG should depart from military reticence rather than risk becoming internationally isolated. From the naval deployment to the Mediterranean in 1987 to the Gulf War to the subsequent missions to halt the wars in former Yugoslavia, a hegemonic project—of Germany's international responsibility—gradually emerged and ultimately succeeded in making possible German participation in multinational military operations. Explicit arguments in favor of out-of-area operations were first discussed extensively in the context of Operation Desert Storm.

In November 1990, following the Iraqi invasion of Kuwait on 2 August 1990 and after months of calls for troop withdrawals, the UN Security Council authorized a US-led coalition to forcibly remove Iraqi troops from Kuwait (Holland 1999; Olsen 2013). Despite the German government's own refusal to send troops, cabinet members defended the mission and demanded that Germany should send troops in the future. Most notably, Chancellor Kohl argued that unification and the emergence of "new dangers for peace and freedom" required that Germany increase its involvement in international peacekeeping; he thus rearticulated the relationship between the use of force and peace. As Kohl claimed, unification "especially reminds us of our duty." Since "freedom and responsibility inseparably belong together," Germans needed to make use of their new-gained freedom in a "responsible" way (Kohl, 12/5, 30 January 1991: 67). Unification required that Germany accept its international responsibility: "With the recovery of our full sovereignty we Germans not only gain in freedom of action but also in responsibility" (Kohl, 12/5, 30 January 1991: 69). Germany should react to the emergence of new dangers neither by "clos[ing] our eyes" nor with "resignation and [an] escape from responsibility," both of which were "dangerous" (Kohl, 12/5, 30 January 1991: 67). Instead, Germany should fulfill its "new role" and "intensify its commitment" to "peacekeeping in the world"—a development that was "correctly expected" by Germany's allies (Kohl, 12/5, 30 January 1991: 69, 90).[23] This demand for Germany to accept its new responsibility was not to advocate "national solo actions or even power ambitions," because "for us there is only one place in this world: in the community of free peoples" (Kohl, 12/5, 30 January 1991: 69). To make such increased engagement in military conflict management possible, Kohl continued, the country needed to "clarify" the "constitutional foundations" regulating the Bundeswehr's scope of operations (Kohl, 12/5, 30 January 1991: 90).

Advocates of out-of-area operations argued that realizing peace and

---

23. See also Zehfuss 2002: ch. 2.

upholding international law under radically changed circumstances required the adaptation of Germany's foreign policy instruments. Since the Cold War, when radical abstinence from any military involvement was considered an appropriate means to avoid war, it had become increasingly questionable whether an exclusive reliance on nonmilitary means was always sufficient, particularly if the perpetrators of violence were unwilling to acquiesce. Thus, advocates of German responsibility clearly articulated an inadequacy (i.e., a lack) in traditional (antimilitarist) German foreign policy, for dealing with current problems. A case in point was Operation Desert Storm, whose purpose, according to Kohl, was "the enforcement of international law and the restoration of peace" against an unresponsive aggressor (12/5, 30 January 1991: 68). The only alternative to Desert Storm, Kohl argued, was an "appeasement policy that puts up with the breach of the law" and that would only "encourage . . . further aggressions" (Kohl, 12/2, 14 January 1991: 23). "Peace at any price" (i.e., appeasement), was unacceptable to Kohl, because a lasting and reliable peace could "only flourish on the basis of freedom, law, and justice" (12/5, 30 January 1991: 69). In Kohl's argument, military reticence, rearticulated as "appeasement policy," which had a particularly negative ring due to its historical connotations (see Adams 1993), was articulated as not equivalent with but contrary to peace, international law, and justice.

This articulation breached the antagonistic frontier established by the old order. During the Cold War, the use of military violence had been firmly linked to the radical Other: military violence was the means of tyrants. In the context of the Iraq war, the use of force was disarticulated from its close connection to the tyrannical Other, and a relation of equivalence with peace was established (see fig. 6). The Bundeswehr, an army of peace during the Cold War because it was never intended to fight, remained such an army after the Cold War, because it was ready to fight in the pursuit of peace. While the general relation of equivalence between the Bundeswehr and peace remained intact, the Bundeswehr's "operational identity" changed.

This emerging project of Germany's new international responsibility gained more and more ground with the unfolding Yugoslav wars, which were seen as prime evidence that an exclusive reliance on nonmilitary instruments of peace policy was doomed to fail. For example, in a 1992 debate about German participation in Operation Maritime Monitor (the naval mission to monitor the arms embargo against Serbia and Montenegro), Foreign Minister Kinkel argued that the Bosnian war had demonstrated

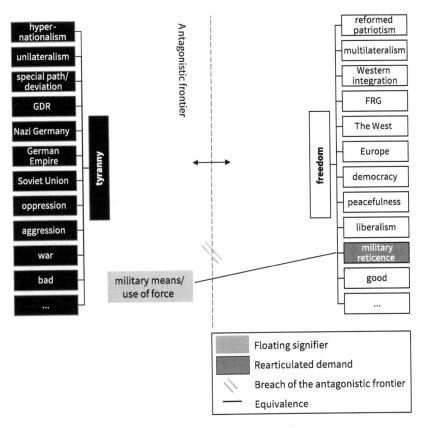

Fig. 6. The breach of the antagonistic frontier

that the traditional instruments of our peace and security policy do not suffice. The conflicts have unfortunately grown faster than instruments to combat them could be developed. . . . *[I]t unfortunately proves to be true that those responsible for violence and aggression only react when it is demonstrated to them that their criminal action leads to a reaction by the international community.* (12/101, 22 July 1992: 8609, italics added).

The articulations in that argument point to a perceived lack of effectiveness in the "traditional," nonmilitary means of German security policy, like diplomacy or economic aid to implement peace. Indirectly, Kinkel's statement tries to grasp the dislocations of the Cold War order: namely, the inability of the old ways of doing security policy to provide a solution for unresponsive aggressive elites. As a result, antimilitarism (or military reticence) became a floating signifier, open to rearticulation.

At the same time, Kinkel already presented a solution: namely, to transcend the "traditional" instruments and to threaten or use military force to demonstrate to the perpetrators of violence that they would have to expect a "reaction of the international community." The project of German responsibility rearticulated the relations between a number of floating signifiers, including peace, military force, war, multilateralism, "alliance solidarity" (Bündnissolidarität), and, more generally, notions of moral righteousness. The use of military force was disarticulated from its close association with the aggressive and tyrannical radical Other and was rearticulated as equivalent with the pursuit of peace, humanitarian demands to help others, support for international law and international organizations, multilateralism, and Western integration, as well as alliance solidarity. Most notably, though, military force was rearticulated as equivalent with the promotion of peace (and the moral good more generally).

Somewhat paradoxically, advocates of German responsibility also managed to incorporate demands for military reticence in an overall argument in favor of military operations. Thus, advocates of German responsibility linked their own demands for military operations to the apparently contrary demand that military means should only be employed as a means of last resort, an ultima ratio, thus strengthening the demand's credibility. For instance, in the debate about Operation Deny Flight, the NATO operation to enforce the no-fly zone over Bosnia and Herzegovina that began in 1993, Foreign Minister Kinkel expressly stated that "military coercion may always only be the very last resort in the keeping of peace," but he continued to point out that "only armed force—and not peace marches—have put an end to Hitler's crimes in 1945. . . . Can we close our eyes to [the fact] that the murder in this world . . . can precisely not be switched off with words alone?" (12/151, 21 April 1993: 12928). Here, in combining references to military reticence with a demand for action, Kinkel utilized the former to justify the latter.[24] German security discourse is ripe with examples for such argument in which references to the limited utility and/or moral undesirability of military force are combined with a demand for precisely such missions. For instance, in the 1992 debate about German participation in Operation Mari-

---

24. In this context, Zehfuss (2007: 233) highlights how comparing Bundeswehr soldiers to the Allied forces that defeated the Nazis silences German atrocities during the Second World War and legitimizes the use of force.

time Monitor, Ulrich Irmer (FDP) claimed that while it was "correct that we cannot reach a problem solution by military means," it was equally true "that there are situations in which, after the exhaustion of all political, all economic, and all diplomatic means, unfortunately no other way exists than, as [an] ultima ratio in a hopeless situation, to grab an aggressor's arm [*einem Aggressor in den Arm zu fallen*]" (12/101, 22 July 1992: 8637).

Similar arguments were made throughout the 1990s with respect to the Serbian leadership and, above all, Serbian/Yugoslav president Slobodan Milošević. Thus, in the parliamentary debate on the German participation in Operation Allied Force, the 1999 NATO intervention, Kinkel argued, "Since the beginning of the Kosovo crisis, we have left no doubt: *the application of violence can only be a possibility as the ultima ratio. But Milosevic remained unreasonable*—despite all the political and diplomatic efforts" (13/248, 16 October 1998: 23128, italics added). Here, Kinkel began by pointing out that military means could only be a means of last resort, thus invoking demands for military reticence and connecting to the mythical notion of a peaceful Self. At the same time, his core demand in the above quote is not to refrain from military operations but the opposite. Milošević is articulated here as "unreasonable." Since he did not respond to anything but force, the West had no choice but to use force; after all, all other options had, in fact, been exhausted. As SPD MP Gernot Erler put it in a different debate on Allied Force, there was "no other alternative" but to "do something" (14/31, 26 March 1999: 2613).

The effect of such articulations is to reinforce the articulation of a benevolent and peaceful Self even if a decision to use violence is made. Since military means can always only be an ultima ratio, the decision to use force implies that all other available means must have already been exhausted. If combined with an argument that "nothing else works," invoking the argument for the ultima ratio only strengthens the demand for military operations. This argument gains additional credibility by drawing on sedimented practices from gender discourse that privilege activity over passivity and strength over (apparent) weakness. Against this background, arguments of no alternative (except "doing nothing") become particularly powerful. In combination with the gendered argument that no alternative to military force exists, antimilitarist arguments can be mobilized to justify intervention—quite contrary to the constraining effect that one would expect antimilitarism to have. Thus, the project of Germany's international responsibility also involved a rearticulation of antimilitarism itself. As a result of this rearticulation, antimilitarism, previously understood as strictly contrary to the use of military force, now allowed for the use of such force as long as all other options had been (or could be said to have been) exhausted.[25]

On the flip side of this articulation, holding on to military reticence in a strict sense was articulated as unrealistic and even immoral under certain circumstances (see, more generally, R. Jackson 2018). For instance, in 1993, CDU/CSU MP Stefan Schwarz argued that military reticence, if taken to the extreme, could amount to "genocidal pacifism" (Schwarz, 12/151, 21 April 1993: 12971). Moral arguments became particularly prominent in the context of Operation Allied Force, the 1999 NATO intervention in Kosovo. For instance, in October 1998, Defense Minister Volker Rühe argued,

> There are enough examples in history that show: it can be immoral to deploy soldiers; *there are, however, also other situations in which one has to say that it is profoundly immoral not to deploy soldiers*, if it is the only chance to stop war and massacre. (13/248, 16 October 1998: 23134)

In Rühe's statement, the use of military violence is articulated as equivalent with, instead of contrary to, notions of moral righteousness, breaching the antagonistic frontier between a peaceful Self and an aggressive past Other. Here again, articulations gain additional clout by drawing on sedimented practices from gender(ed) discourses. Arguments demanding help for an allegedly "defenseless" civilian population, as SPD chairman Rudolf Scharping put it in debates about Bosnia (13/48, 30 June 1995: 3959), draw on sedimented practices portraying the non-Western Other as in need of Western help. Such arguments seem more credible simply by virtue of their familiarity, independent of whether they are the most "rational" option in terms of expected utility or potential unintended negative consequences (Chandra 2013).

These articulations received additional strength because they drew on established discursive patterns from gender discourse, such as the logic of masculinist protection (Young 2003). Thus, advocates of military operations often invoked the dangers or difficulties of any given mission and highlighted the selfless service of the soldiers, for which citizens (including mem-

---

25. This shift is quite similar to the changing meaning of multilateralism that Baumann (2006) discussed, and it supports Junk and Daase's (2013) argument that acceptance of military operations depends on exactly how they are framed.

bers of the Bundestag) should show their gratitude by supporting them (to be expressed by a broad vote of support in the Bundestag). For instance, in the run-up to Operation Allied Force, Defense Minister Rühe claimed,

> The German soldiers need, for their difficult mission, the unconditional and visible backing of the German parliament. . . . Do not underestimate what the vote of the German Bundestag means to our soldiers! . . . The difficult situation [of the soldiers] is much easier to bear if one senses [that] [t]he representatives of the German people are behind our soldiers. (Rühe, 13/248, 16 October 1998: 23134)

Rühe's claim illustrates, in particular, how gratitude owed to the soldiers is inextricably linked to support for the government's policies. The claim here is that support for the decision would make the difficult mission "much easier to bear." Vice versa, failure to support the government's motion equals being ungrateful to the soldiers who, after all, risk their lives for "us." Despite Rühe's suggestion, what soldiers might actually think of any specific mission is not considered, because they are never asked in the first place.

At the same time, the lesson to be drawn from past violence was rearticulated, from a rejection of military involvement abroad to a duty not to stand aside.[26] For instance, in 1999, Defense Minister Scharping stressed an "obligation due to the experiences of the first half of this century" (14/30, 25 March 1999: 2424). In this context, Serbian atrocities of the 1990s, on one hand, and German war crimes during the Second World War and the Holocaust, on the other, were articulated as equivalent (although not equal). For instance, in April 1999, chancellor Gerhard Schröder emphasized the need to avoid "that a part of Europe lapses into oppression and barbarity" (14/35, 22 April 1999: 2762). Similarly, Foreign Minister Joschka Fischer claimed that Serbian leader Slobodan Milošević pursued "a policy of violence and of the past" (14/25, 22 April 1999: 2777).[27] Thus, rather than as a new antagonistic Other, the perpetrators of the new conflicts were articulated as instantiations of the tyrannical Other.

---

26. In particular, originally, the deployment of German soldiers to regions the Wehrmacht had occupied had been firmly rejected by the Kohl government. This rejection is commonly referred to as the "Kohl doctrine" (von Krause 2013: 199–200; Zehfuss 2002: 129).

27. Fischer (1999) put this argument in even clearer terms during the extraordinary party conference of the Greens on 13 May 1999, where he directly invoked Auschwitz.

At the same time, "war" remained closely connected with the radical Other.[28] While policymakers demanded that the use of military violence be made possible, they simultaneously insisted that "German foreign policy was and also in the future remains peace policy" (Kinkel, 12/151, 21 April 1993: 12929) and that the Bundeswehr's only purpose remained to serve peace. For example, in a debate about Operation Maritime Monitor, Kinkel insisted that "all" German soldiers, "together and entirely without qualification, serve peace and nothing else. . . . No one should insinuate that the Bundeswehr and the soldiers of the Bundeswehr serve any other goal and purpose than the law and peace" (12/101, 22 July 1992: 8619). The Bundeswehr that had been an army of peace during the 1980s by virtue of it not being intended to fight remained an army of peace in the 1990s because it was ready to fight for peace.

In addition, other demands were incorporated into the chain of equivalences, most notably the demand for continued integration in NATO, the EU, and the international community. Not only were out-of-area operations legitimized with the need for solidarity within NATO, but such solidarity was said to be a precondition for Germany not becoming isolated in the international community as such.[29] For instance, in the debate about Operation Deny Flight in April 1993, Foreign Minister Kinkel argued that a new consensus was needed.

> Today it is . . . about a question of fundamental foreign and security political significance, namely: Will we find, as a united and sovereign Germany after the end of the bipolar world of the East-West conflict, . . . a new foreign and security political consensus that makes us, in a changed world situation, into a *responsible actor of the world community* that is capable to act? *Are we willing to assume without reservation the peace tasks that the community of peoples expects from us* as a leading industrial nation in the face of entirely new security political challenges? (Kinkel, 12/151, 21 April 1993: 12925)

---

28. This connection arguably contributed to the difficulties of acknowledging German soldiers' claims, many years later, that war was going on in Afghanistan; war was considered an act of the Other, not of the new Germany.

29. The solidarity argument was not entirely new at the time. Already in 1964, when a potential German participation in a NATO force to prevent a war between Turkey and Greece over Cyprus was under debate, then Foreign Minister Gerhard Schröder (CDU) argued that for "alliance reasons" only, Germany could not abstain (quoted in Troche 2000: 189). This emphasis on alliance solidarity was shared among representatives of the SPD, most notably Herbert Wehner. Bundestag vice president Carlo Schmidt (SPD) even argued that the Basic Law could be ignored in this case, as such a "police action" could not be seen as a war (quoted in Troche 2000: 189).

While participation in military operations contributed to Germany's international integration and made the country capable to act in a changed world, failure to live up to international expectations would render Germany "incapable to function as an alliance partner" (*bündnisunfähig*) and would lead to the country's "isolation . . . in the community of states" (Kinkel, 12/151, 21 April 1993: 12928, 12929).[30] Alliance solidarity, the capability to function as an alliance partner (*Bündnisfähigkeit*), and "reliability" more generally were prominent demands in the debates about the Yugoslav wars.[31] Similarly, support for international law was said to require participation in military operations. Already in his speeches on Operation Desert Storm in 1991, Chancellor Kohl argued that the "community of states" could not "tolerate this breach of law" (12/2, 14 January 1991: 2).[32]

The project of Germany's international responsibility also rearticulated the proper "lessons" to be drawn from Germany's past (see Zehfuss 2007: 7–10). Thus, in a debate about the Gulf War, Chancellor Kohl claimed that "our own history teaches" us "that right must never give way to wrong"[33] and that "aggressors have to be confronted in good time" (12/2, 14 January 1991: 22). In a quite similar vein, in the debates about former Yugoslavia a few years later, Foreign Minister Kinkel argued that

> the experience of our most recent history, namely, the period of National Socialist dictatorship . . . and the illegitimate regime [*Unrechtsregime*] of a different kind in the former GDR, *really justifies a special duty for German foreign policy to actively stand up for peace and human rights internationally*. (12/101, 22 July 1992: 8609, italics added)

---

30. Thus, German decision makers themselves constructed what Markus Kaim (2007) has called the "multilateralism trap," as also observed by members of the PDS / Linke Liste at the time. Andrea Lederer argued that "the federal government and the parliamentary groups supporting it have conducted a foreign policy that has provoked the expectations" that were then invoked to justify military operations (12/151, 21 April 1993: 12942).

31. See, for instance, Glos, 13/48, 30 June 1995: 3984; Kinkel, 13/48, 30 June 1995: 3957; Rühe, 13/87, 9 February 1996: 7679; Volmer, 13/248, 16 October 1998: 23151; Kinkel, 13/248, 16 October 1998: 23129. These arguments have been discussed in detail by Schwab-Trapp (2002), and the importance of arguments about alliance solidarity is well established in the literature on out-of-area operations.

32. See also the unfolding debates in 12/2, 14 January 1991; 12/3, 17 January 1991; and 12/5, 30 January 1991.

33. The original German expression juxtaposes *Recht* with *Unrecht*, which can be translated "wrong"/"right," "justice"/"injustice," or "law"/"lawlessness." In that sense, the original German formulation includes both aspects of moral righteousness and legality.

The true lesson of history, then, was not to reject the use of military force once and for all but, rather, to never again stray from Western Europe's common path. Fulfilling the demand to draw the right lessons from history meant not that one should fundamentally oppose military violence under all circumstances but, rather, that one should avoid any new special paths. As Foreign Minister Kinkel claimed, "One lesson from this [Germany's] history can really only be: never again sheer off from the community of Western peoples, never again special paths, [not even] that of moral know-it-all attitude and of ethics of conviction" (12/151, 21 April 1993: 12928).

Even the constitutional legality of military operations was transformed during the 1990s, if only through multiple discursive interventions by the BVerfG. Constitutionality was a particularly important bone of contention with respect to UNOSOM II, the UN mission in Somalia, and Operation Deny Flight in Bosnia. In the case of Somalia, much of the debate circled around whether the operation was humanitarian or military in nature. Before the authoritative ruling of the BVerfG in 1994, the legal consensus was that purely humanitarian missions such as relief aid were constitutional, whereas everything else was considered a violation of the Basic Law. Not surprisingly, advocates of German participation in UNOSOM II argued that the mission was a "humanitarian relief action" (Schäuble, 12/163, 17 June 1993: 12996), while opponents claimed that it was an "obviously dangerous military operation" and, consequently, "unconstitutional" (Fuchs, 12/163, 17 June 1993: 12995, 12994). The debates culminated in an appeal by the SPD parliamentary group to the BVerfG, which was dismissed (BVerfGE 89, 38).

After that, the debate shifted toward the desirability of UNOSOM II and combat operations in general. Politically, the SPD opposed peace enforcement operations because of a doubt that military force was suitable as an instrument to achieve peace. For instance, on 24 June 1993, one day after the BVerfG's dismissal of the SPD parliamentary group's appeal, SPD federal manager Günter Verheugen repeated that the party was "not willing . . . to permit combat missions outside of alliance and national defense, because we do not believe that the problems in the world can be solved through the use of military violence" (12/166, 24 June 1993: 14332). While CDU/CSU MPs articulated military operations and peace as equivalent, SPD MPs stressed the two elements' differential character, putting in doubt that they went hand in hand.

Even more controversial than the Somalia operation was the deployment of German air force members as part of the AWACS crews in Operation

Deny Flight.[34] The central question was whether the mission required a constitutional amendment. Not being able to find consensus, the parliamentary groups of the SPD and FDP (the latter a member of the ruling coalition) turned to the BVerfG on 8 April 1993, to determine, via an interim measure, whether German soldiers participating in AWACS flights was constitutional. In addition, the SPD parliamentary group appealed to the court to decide whether the federal government's deployment of troops to participate in UNOSOM II was constitutional (Baumann and Hellmann 2001: 73-74; Philippi 1997: 48).[35] In 1994, the BVerfG ruled that out-of-area operations were permitted if conducted within systems of collective security such as the UN and if endowed with a constitutive Bundestag mandate (BVerfGE 90, 286), articulating out-of-area operations as principally equivalent with the demand for constitutional legality. Given the undisputed authority of the BVerfG on constitutional matters, the ruling ended the debate in favor of the ruling coalition. This example clearly illustrates, first, that from a discourse theoretical perspective, legality is not an objective fact but is established through the performative act of a court and, second, that discourse is not a level playing field, for due to sedimented discursive practices, the BVerfG alone has the "interpretive power" (*Deutungsmacht*) to declare any demand equivalent with or contrary to the provisions of the Basic Law (Brodocz 2009: ch. 5).[36]

Another factor that contributed to the project's success was that it managed to provide an empty signifier that functioned as a surface of inscription for individual demands. This signifier was the demand for responsibility, or, to be more precise, for Germany to accept and live up to its purported new international responsibility. As the above quotes illustrate, responsibility featured prominently in the debate.[37] Already in the context of the Gulf War, Chancellor Kohl had called for Germany to accept its international responsibility, and the demand was echoed across time and party lines. For instance,

---

34. See 12/150, 26 March 1993: 12867–81; 12/151, 21 April 1993: 12925–73.

35. As Philippi (1997: 48) points out, the BVerfG does not offer legal opinions, so the SPD and FDP had to turn to the court on the basis of a dispute between different constitutional bodies (*Organstreitverfahren*). The purpose of such proceedings is normally to determine which constitutional body can legitimately decide on a certain issue, but it was used here to let the court somewhat indirectly rule on the constitutionality of out-of-area deployments.

36. This uncontested authority is itself the result of past discursive struggles (Brodocz 2009: 145–46, ch. 6; Vorländer 2006: 9).

37. On responsibility in this context, see Crossley-Frolick 2017; Geis and Pfeifer 2017; Pfeifer and Spandler 2014; Stahl 2017; Stark Urrestarazu 2015: 182–83.

SPD MP Walter Kolbow cited Germany's "international responsibility" as the main reason for the need to participate in (traditional) UN peacekeeping operations (12/70, 16 January 1992: 5884), and members of the Greens pointed to Germany's "international responsibility" in the context of the Yugoslav wars (Wollenberger, 12/101, 22 July 1992: 8630). The notion of responsibility continues to permeate foreign policy debates to this day, as illustrated by federal president Gauck's (2014) much-noted speech at the 2014 Munich Security Conference and by a more recent *Foreign Affairs* article by Foreign Minister Frank-Walter Steinmeier (2016).

Remaining particular content made responsibility particularly effective. No empty signifier is ever completely empty, and because of its elevated position, the empty signifier influences the overall character of the hegemonic project it symbolizes. In the case of responsibility, what matters most is the strong notion of duty associated with responsibility or, more precisely, the German term *Verantwortung*.[38] According to the *Duden*, the most prominent dictionary of the German language, *Verantwortung* commonly refers to someone's "obligation [*Verpflichtung*] to ensure that (within a specific framework) . . . the right thing is being done" (*Duden* 2019).[39] As is the case with the word *duty* in the English language, the German *Verpflichtung* and the associated verb *verpflichten* (to oblige someone) refer to a binding commitment, something that must be done (*Duden* 2019). Rendering military missions in terms of *Verantwortung* means that there is no choice to be had. What makes a duty dutiful is precisely that it is not a decision between two equally valid (contingent) options. Regardless of whether the addressee of a duty, obligation, or responsibility wants to fulfill it, the duty will have to be carried out anyway. The subject faces a decision only between accepting the duty or choosing not to fulfill it, at the risk of being regarded as irresponsible and unreliable. Because established social practices demand that one fulfill one's duty (to family members and friends, to one's employer, etc.), it is arguably much more difficult to object to a demand when it is articulated as a

---

38. The discourse does not necessarily always use the term *Verantwortung*. At times, decision makers point to Germany's "obligations" (*Verpflichtungen*) (e.g., Hoyer, 12/53, 6 November 1991: 4399; Dregger, 12/6, 31 January 1991: 112; Kohl, 12/5, 30 January 1991: 98), "duty" (*Pflicht*) (e.g., Zierer, 12/70, 16 January 1992: 5886), or the allies' "right" (*Anspruch*) to German "solidarity" (Kohl, 12/2, 14 January 1991: 22).

39. Similarly, the English word *responsibility* refers to "a duty to take care of something," and the word *duty*, in turn, refers to "something that you *have to do* because it is part of your job, or something that you feel is *the right thing to do*" (*Cambridge Academic Content Dictionary* 2014, italics added).

matter of responsibility rather than choice. In the context of foreign policy, being perceived as unreliable carries the additional risk of becoming internationally isolated, which reinforces the overall power of the project.

Despite its success in making the use of military violence possible, the project of responsibility remained limited in a number of respects. Most notably, the project did not, at the level of the overall security discourse, establish a new security order. Quite to the contrary, the old order remained largely in place, if in a partially rearticulated and generally dislocated version, with traditional territorial defense being the main purpose of the armed forces. Moreover, the project did not yet establish out-of-area operations as a taken-for-granted, "normal" social practice. Instead, each individual operation was justified on its specific grounds, and no overarching discursive formation emerged to justify these missions as a social practice. Importantly, out-of-area operations were largely defended with arguments other than German security, most notably international expectations and humanitarian demands—that is, responsibility not to the German people but to Germany's allies, the international community, normative principles, and, linked to that, (a reinterpretation of) the "lessons of history."

Moreover, the project did not establish a new antagonistic frontier. Rather, the opponents in specific missions (Iraqi president Saddam Hussein, Milošević, etc.) were articulated as instantiations of the tyrannical Other and as equivalent, in that sense, with Germany's own past. Thus, while the project was successful in making military operations possible, it did not establish an overall framework that provided an authoritative description of the post–Cold War security situation. It neither outlined what German security policy should look like under changed circumstances nor defined the role of the armed forces within such a security strategy. Put in more conventional terms, the project did not (yet) formulate a new German grand strategy.

CONCLUSION

This chapter has outlined how the project of responsibility managed to make out-of-area operations possible. The project's success can mainly be explained by three factors: (1) the articulation of military operations as equivalent with a broad range of demands, including the promotion of peace as well as antimilitarist demands; (2) the incorporation of sedimented discursive practices, including the myth of the peaceable nation as well as established discursive patterns surrounding notions of masculinity/femininity and modernity/coloniality; and (3) the construction of responsibility as an empty signifier, as a result of which military operations appeared not so much as a matter of choice but as a duty to be fulfilled. As a result of the discursive rearticulation of the Cold War order, the Bundeswehr's identity as an "army of peace" was transformed. During the Cold War, the Bundeswehr had been an army of peace because its readiness guaranteed that Germans would not have to fight in the first place; during the 1990s, it stayed an army of peace because it was actually willing to fight for peace.

At the same time, the project remained limited. Most notably, while the project was successful in making out-of-area operations possible, it did not establish a new security order. Individual out-of-area operations were justified mainly on the basis of international expectations and humanitarian arguments, but military operations were not yet established as a social practice within a broader security order that provided an overall rationale for why these operations were needed for German security. While military operations were not considered a taboo anymore, they had not become a matter of routine either. Above all, the project did not formulate a new grand strategy.

This situation began to change with the emergence of what would become the project of networked security, which, for the first time, bound the political practices of the 1990s, some of them originally articulated as contrary, into a common hegemonic project that provided an overarching representation of the security situation after the end of the Cold War and that spelled out a strategy of how to maintain security in a changed environment. In the late 1990s, advocates of what was then still called "comprehensive security" argued that the German policy of conflict prevention should pursue a whole-of-government approach, combining civilian and military means. Emerging out of this project, after 9/11, was a new security order that clearly formulated a new German grand strategy. This establishment of a new security order and of military operations as an integral part of a "modern" security policy is the topic of the two following chapters.

CHAPTER 4

Peace in Europe

Comprehensive Security as a New Paradigm for German Policy on Conflict Prevention

This chapter examines the emergence of comprehensive security as a hegemonic project aiming to install a new order in the German discourse on conflict prevention. In the previous chapter, I examined how floating signifiers (most notably military operations and antimilitarism) that had been part of the old security order became rearticulated, enabling German participation in out-of-area operations. At the same time, the scope of the discourse of Germany's international responsibility was also limited to making out-of-area operations possible. It did not provide a new overarching framework for German policy on conflict prevention,[1] let alone German security policy as a whole. Put in more conventional terms, the discourse of Germany's international responsibility did not develop a new grand strategy, the purpose with which security orders are essentially concerned.

This broader process of situating out-of-area operations within a new grand strategy is the concern of the project of networked security, which I address in more detail in chapter 5. The present chapter examines the project of comprehensive security, out of which the project of networked security grew. The project of comprehensive security originally emerged not with the aim to implement a new security order (and, with it, a new grand strategy) but as a more modest attempt to provide a rationale for German policy on conflict prevention, security policy's smaller subfield concerned with the prevention of violent conflict abroad. In this context, the demand for com-

---

1. In the literature, the policy on conflict prevention is often referred to as peace policy (*Friedenspolitik*), often however without a clear definition of what peace policy is and what separates it from, e.g., security policy (Egbering 2011; Geis 2008; Gießmann 2004; Schlotter 2008).

prehensive (later, networked) security—or, more precisely, for the adoption of a comprehensive security policy—refers, very generally, to the need to combine the various military and nonmilitary instruments of different state and nonstate actors into a whole-of-government approach (see Stengel and Weller 2010). After 9/11, this template, originally formulated just for policy on conflict prevention, was expanded and promoted as a candidate for a new German grand strategy.

To avoid any potential confusion, I here use the term *comprehensive security* to refer to the project to reorganize German policy on conflict prevention. I reserve the term *networked security* for the project that emerged after 9/11 and targeted the much broader security discourse (see table 9). The project of comprehensive security strove to establish comprehensive security as the overarching framework for German policy on conflict prevention, not for security policy as a whole, which, during the early and mid-1990s, was still articulated mainly in traditional military terms.

The distinction between the project of comprehensive security and that of networked security is a heuristic distinction that does not claim to be an "objective" representation of the parliamentary discourse (insofar as there could ever be one). Instead, its purpose is to reduce complexity and thus make it easier to follow the argument. Nevertheless, my usage of the terms is not exactly congruent with the way the terms are used by discourse participants. Both projects were formulated under the SPD/Green government of Chancellor Schröder, whose members consistently used the term *comprehensive security* to refer to the strategy for, first, conflict prevention and, later, security policy as a whole. The term *networked security* was introduced only after the change of government in which Chancellor Angela Merkel (CDU) came to power, no earlier than 2006, in the context of the white paper on defense published that year.[2] Thus, in the following discussion, I use a term that was introduced in 2006 to describe a project that emerged in 2001.

TABLE 9. Comprehensive versus networked security

Discourse	Extension/competence	Time frame (broadly)
Comprehensive security	Conflict prevention	Late 1990s and early 2000s
Networked security	Grand strategy / security policy in general (including conflict prevention)	Early 2000s onward

2. To my knowledge, Defense Minister Franz Josef Jung (16/51, 21 September 2006: 4974) began to use the notion of a "networked security policy" in autumn 2006.

Although this chapter is primarily concerned with the late 1990s and early 2000s, I here sometimes cite speeches from before that period. The reason is that the project of comprehensive security (and, by extension, its successor) took up articulations that had already been present in the discourse. This adoption partly explains the project's broad appeal. It managed to incorporate demands of critics and champions of out-of-area operations alike and to rearticulate them into a common project.

This chapter examines the emergence of the project of comprehensive security, while the expansion of the project to encompass the entire security discourse is the subject of chapter 5. The first section of the present chapter offers a detailed discussion of the construction of a chain of equivalent demands. The second section details the construction of comprehensive security as an empty signifier. The third section takes a broader perspective and analyzes the formation of an antagonistic frontier at the level of the overall discourse.

LINKING DEMANDS IN/TO CONFLICT PREVENTION

The emergence of the project of comprehensive security coincides broadly with the election victory of the SPD/Green coalition in 1998. After that victory, coalition members more forcefully argued for the need to rethink conflict prevention or, as it was called in the discourse, "crisis prevention" (e.g., Schröder, 14/35, 22 April 1999: 5764). In terms of the demands incorporated into the project of comprehensive security, one can distinguish between

1. the articulation of substantial demands as equivalent, either (a) for policy goals or even whole policy fields (e.g., environmental, peace, or development policy) or (b) for certain instruments of foreign and security policy (e.g., military operations, diplomatic negotiations, or development aid), and
2. the incorporation of different individual and collective subjects (i.e., actors) into a chain of equivalence (e.g., German soldiers, the German or other populations, the FRG, the US, different articulations of Europe, the UN, or the "international community" as such).

The project of comprehensive security emerged as a response to a persistent lack of peace in the world. This intent is most clearly visible in Chancellor Schröder's inaugural government policy statement in November 1998.

Berlin is . . . also the city that had for decades been split by the East-West conflict. As happy as we Germans are about it [the Cold War] having been overcome, as aware are we also [of the fact] that the end of the Cold War has not
by a long shot [*noch lange nicht*] brought world peace.

The world political upheaval has triggered new instabilities and violent
conflicts in many regions, also on our doorstep in Europe. [The] misery of
refugees, scarcity of resources, and ecological destruction in the countries of
the South are dangerous breeding grounds for these and new conflicts.

In light of such risks, but above all in light of the opportunities for international cooperation, the world expects of us more than ever that we do justice
to our obligations within the framework of our alliances. We remain reliable
partners in Europe and in the world. (Schröder, 14/3, 10 November 1998: 63)

Here, Schröder clearly identified a persistent lack of "world peace" even after
the end of the Cold War. This lack of peace was due to what Schröder called
"new instabilities and . . . conflicts," which, in turn, were supported by the
suffering of refugees, resource scarcity, and environmental destruction.

Already in Schröder's statement, one can see that antagonism and equivalence are practically inseparable. Because these problems were a challenge
to stability not just for Germany but for "Europe" and "the world" as a whole,
these different collective subjects were (or at least should be) united in a
common project to combat them. The above quote from Schröder illustrates
that in the case of the project of comprehensive security, the articulation of
different substantial demands as equivalent takes place mainly through the
articulation of different policy problems as linked. Schröder articulated
armed conflicts—or "crises," as they were broadly referred to by members of
the Schröder government (Scharping, 14/3, 10 November 1998: 113)[3]—as
being caused or aggravated by other phenomena: namely, the "misery of
refugees," resource scarcity, and environmental problems. As contributing
factors to armed conflict, these were constructed as equivalent ($\triangleq$) with the
latter:

instabilities/violent conflicts $\triangleq$ misery of refugees $\triangleq$ ecological
destruction $\triangleq$ resource scarcity

---

3. In the context of the German security discourse of the 1990s, *crisis* functions as
somewhat of a catch-all term broadly referring to any (risk of the) occurrence of mass
violence, including armed conflict, large-scale human rights violations, etc. (see, e.g.,
Federal Government 2004).

Rather than as separate issues, these problems were articulated as linked to armed conflict. With respect to the demand for (world) peace, then, the misery of refugees, resource scarcity, and environmental problems were equally articulated as contrary to the Self, as it is clearly implied that "we" want peace.

Similarly, in a debate on 10 November 1998 (following the first government policy statement of Chancellor Schröder cited above), Defense Minister Scharping (SPD) claimed that in light of "altered and new challenges for our security and for that of our continent," German security had to orient itself toward the "goal of comprehensive security," which had to "include also causes of crises like hunger, underdevelopment, terror, and hatred between population groups" (14/3, 10 November 1998: 114). Here, Scharping provided a similar yet slightly different articulation, listing hunger, underdevelopment, and ethnic hatred ("hatred between population groups") as causes of crises (like armed conflicts):

crises/armed conflict $\triangleq$ hunger $\triangleq$ underdevelopment $\triangleq$ terror $\triangleq$ ethnic hatred

This chain of equivalent problems (to stay with the discourse theoretical terminology) became gradually widened, with decision makers sometimes leaving out some phenomena but mainly adding new ones.

Authoritarian (i.e., nondemocratic) rule was also articulated as causally linked to armed conflict (incorporating elements from liberal discourse more generally as well as articulations of the Soviet Union during the Cold War as aggressive by virtue of authoritarianism). In a debate about NATO's 50th anniversary in April 1999, Chancellor Schröder argued, "We have seen: the danger of armed conflicts and militant disputes exists above all where democracy is lacking and where dictators want to force their will on their peoples and behave accordingly" (14/35, 22 April 1999: 2762). Here, authoritarian rule as well was incorporated into the (emerging) chain of equivalence responsible for the continued lack of (world) peace. Thus, to rid the world of armed conflict, one needed to simultaneously address authoritarianism as a problem.

These examples equally illustrate how individual articulations or texts accumulate to discursive formations on the level of the overall discourse. For instance, viewed in isolation from each other, the above articulations by Schröder and Scharping emphasized a clearly specified number of certain phenomena while leaving out others, and both differed from each other. On the level of each individual articulation, we can clearly distinguish which phenomena were considered part of the category of the new challenges. Scharping connected armed conflict to hunger, underdevelopment, and environmental problems, while Schröder articulated (in the 1999 quote above) a connection between armed conflict and authoritarianism. When one reads these statements together, however, a different, wider picture of the new challenges emerges, to include the entirety of the phenomena cited by both Schröder and Scharping:[4]

crises/instabilities/armed conflict $\triangleq$ authoritarianism $\triangleq$ misery of refugees $\triangleq$ ecological destruction $\triangleq$ resource scarcity $\triangleq$ underdevelopment $\triangleq$ ethnic hatred $\triangleq$ hunger

On the level of these individual statements, only selected policy problems are articulated as linked, but emerging on the level of the discourse as a whole is an image of a common cluster of interlinked and inseparable (i.e., equivalent) problems. Rather than being distinct phenomena, these new threats (or "new challenges," as they were then mainly called: see Schröder, 14/35, 22 April 1999: 2764) required that they be solved together or not at all. Moreover, all these phenomena together were to be blamed for the continued lack of world peace. This implies the assumption that if these new threats (armed conflicts and their causes) were to be overcome, world peace would materialize. To cluster problems in this way is to articulate a set of phenomena into a common radical Other that is preventing the achievement of a fully constituted, peaceful identity for the world community. The important point about the new threats is that they were articulated as equivalent, as a common totality of causally linked problems.[5] This articulation is clearly

---

4. It is difficult to clearly distinguish between these phenomena in practice. For instance, poverty and hunger are so closely related that it would be problematic to systematically separate them as "variables" in a quantitative study.

5. When I speak about causal connections here, I claim not that they exist objectively but that the different phenomena were articulated as causally linked in discourse. What I observe is a truth claim regarding a causal connection between, say, state failure and terrorism (see Schlichte 2008). This understanding is closer to the notion of causal beliefs or ideas as "beliefs about cause-effect relationships" (Goldstein and Keohane 1993a: 10). These beliefs need not be factually correct; what matters is that they are widely believed. One example is the still widespread belief that photographic images display reality in an unmediated fashion, which prevails despite the fact that a photograph can always only show some limited frame of reality—let alone the possibilities for fabricating photographic images in the age of Photoshop (see Pettersson 2011; Shim 2014).

visible, for example, in descriptions of environmental problems and other problems as the "breeding grounds" for armed conflict (Schröder, 14/3, 10 November 1998: 63) or in claims that the risk for conflict is higher where democratic governance is missing.

The flip side of this articulation of interlinked problems is the articulation of a chain of equivalent demands that, with respect to the goal to overcome the lack (of peace), were articulated as contrary to the new threats. The articulation of a number of different policy problems as a common Other means that the corresponding social demands (to address the problems) emerge as equivalent. For instance, if armed conflict and a lack of democratic governance are inherently linked, demands for conflict prevention and demands for democratization cannot be separated anymore. Conflict prevention requires democratization, and vice versa.

This articulation of different demands as equivalent is clearly visible in the discourse. Thus, coalition members claimed that peace and development could not be understood as disparate issues anymore (much less as contrary ones—e.g., in the competition for scarce government resources). For instance, Chancellor Schröder argued in April 1999 that after the end of the Cold War, "security *can less and less be achieved by military means alone. A modern security policy has to think about peace and economic-social development together*"; he concluded that "efficient crisis management and effective crisis prevention" required a comprehensive perspective (14/35, 22 April 1999: 5764, italics added). Because underdevelopment, a lack of democratic governance, and armed conflict were linked, a "modern" security policy had to regard peace and development as inseparable.

Three additional aspects are relevant here. First, Schröder's response to the dislocations of the Cold War order was political in nature: he challenged the Cold War order's narrow military focus and articulated the end of the Cold War as a central turning point, demanding policy change. In doing so, Schröder also took up the demands formulated by critics of the Kohl government's security policy. These critics argued that the end of the Cold War meant that the relative importance of military security provision was significantly decreased (see the discussion in chapter 3). At the same time, however, Schröder did not reiterate demands for disarmament, which ensured the project's ability to connect with demands by the CDU/CSU and the FDP too. In this way, previously contrary demands emerged, if implicitly, as equivalent.

Second, Schröder also formulated an alternative to the old order, capable to overcome the latter's dislocations. He claimed that a "modern," comprehensive security policy could overcome the dislocations of the old order by broadening the perspective of security policy, from a traditional focus on military means, to "think[ing] about peace and economic-social development together." In this way, a modern security policy would be able to domesticate what the old order could not (see Volmer, 14/12, 4 December 1998: 724).[6]

Third, a "modern" (i.e., comprehensive) security policy—as opposed to what Ludger Volmer (Greens), minister of state in the Foreign Office, called the CDU/CSU's "traditional" one (14/12, 4 December 1998: 723)—did not understand "peace" and "economic-social development" as disparate demands but, instead, required that the two were understood as going hand in hand. In this way, development policy was incorporated into policy on conflict prevention, anticipating the explicit articulation of a "security-development nexus" in the German security discourse (see Duffield 2001; Stern and Öjendal 2010; Ziai 2010). This connection was even more clearly stated by Schröder in his inaugural government policy statement in November 1998, in which he claimed,

> We know from [our] own experience: peace needs economic development, and economic development needs peace. Crises can only be solved permanently where people realize that peace and democracy are profitable and that peaceful development noticeably improves their situation. (14/3, 10 November 1998: 65)

Articulated as equivalent here are not only peace and economic development but also democracy. Simply through enumeration, democracy was implicitly claimed to go hand in hand with peace, a claim drawing on sedimented discursive practices, in liberal political discourse, that linked peace and democracy (Neocleous 2011: 93–94). Although this claim is factual, the articulation of equivalence between development, democracy, and peace is far from self-evident. The relationship between democracy and democratization, on one hand, and peace, on the other, remains controversially discussed in the academic literature (see Geis 2001, 2013b; Geis and Wagner

---

6. Although decision makers speak about security policy, the threats to German security at the time were not the issue. Despite multiple references to security policy, the debate was not concerned with grand strategy in a narrow sense. At that point in time, the project was not (yet) about a reformulation of security policy as a whole but was limited to the narrower field of conflict prevention.

2011; Mansfield and Snyder 1995, 2002, 2009; Müller 2004a), as does the link between economic development—certainly if it entails the spread of capitalism (e.g., Schneider 2017; Siegelberg 1994)—and peace (see Anderson 1999).

In addition, further demands and entire policy fields were linked to conflict prevention. For instance, foreign economic policy was enlisted for the purpose of fostering peace, development, and democratization. Schröder stated in 1998, "Also our foreign economic relations are to serve peace and democratization" (14/3, 10 November 1998: 65). In a debate on the occasion of NATO's 50th anniversary, Defense Minister Scharping argued that NATO was "characterized by a supremely modern and comprehensive understanding of security. . . . Economic recovery, domestic political stability, and external security belong inextricably together not only in the eyes of the founding fathers" (14/35, 22 April 1999: 2770). Equally, at least in debates about specific operations, immigration policy was linked to peace policy, with policymakers citing "refugee wave[s]" (Kinkel, 12/101, 22 July 1992: 8610; Kolbow, 14/45, 17 June 1999: 3732) as consequences of armed conflict.

The result of these articulations was a broad chain of equivalent demands, incorporating various policy goals (peace, development, democratization, human rights) and policy fields (e.g., peace, development, and foreign economic and environmental policy) under the broad umbrella of a comprehensive security policy:

conflict prevention ≙ development policy ≙ environmental protection ≙ foreign economic policy ≙ human rights policy ≙ democracy promotion ≙ internal stability ≙ immigration control

Within the project of comprehensive security, otherwise disparate demands like environmental protection and peace promotion were rearticulated as actually inherently linked and mutually dependent. The project also incorporated a number of demands that had been connected only marginally to the previous security discourse (see fig. 7).

The project of comprehensive security involved the deliberate breach of the super-differential boundary that previously separated conflict prevention from other thematic discourses, such as the development and environmental discourses. This incorporation significantly increases the reach of the hegemonic project, by establishing links between previously largely separate discourses. It bolsters the project's claim to not just advocate a set of

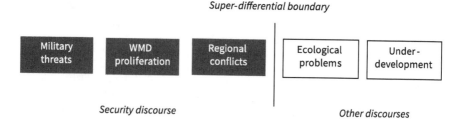

Fig. 7. The security discourse before the project of comprehensive security

particular demands but promote the general common good. Equally, it makes the project more credible, because the project then resonates with a broad range of sedimented demands.

The project of comprehensive security incorporated a number of articulations that emerged as part of the project of Germany's international responsibility, most notably the articulation of military operations as equivalent, if only as a means of last resort, with a number of goals, including peace and development. For instance, in his November 1998 government policy declaration, Chancellor Schröder claimed that under his leadership, "German foreign policy is and remains peace policy," which "explicitly" included the "willingness to contribute to peacekeeping and peace-preserving measures and missions," and that the Bundeswehr remained "an army that serves the peace" (14/3, 10 November 1998: 64). This formulation of the Bundeswehr took up understandings prominent among conservatives and liberals during the Cold War and increased the project's credibility for those audiences. It incorporated the modified articulation of the Bundeswehr as an army of peace, according to which it was an instrument of peace by virtue of its readiness to fight for it (see chapter 3). At the same time, the project also integrated the argument of ultima ratio, according to which military means could only be a means of last resort. For example, in November 1998, Schröder argued that "the keeping available [das Vorhalten] of military capabilities" should "serve crisis prevention" and that "its application has to remain the ultima ratio of peace policy" (14/3, 10 November 1998: 64). Foreign Minister Joschka Fischer (Greens) similarly stressed the importance of adhering to "a policy of self-restraint" (14/3, 10 November 1998: 108). When military deployments were deemed necessary, they were generally said to be without alternative. For instance, regarding Allied Force, Fischer, drawing on gendered constructions again for credibility, argued that the choice was between military intervention and "a bending, a ducking away" from Milošević, which would only lead to "even more displacement and even more destruction" (14/35, 22 April 1999: 2778).

This articulation of military force as without alternative (because all other options had been exhausted) was further reinforced by German decision makers' insistence on their personal reluctance in making a deployment decision. This reluctance had already been a prominent feature of the out-of-area debate of the 1990s. For example, in June 1995, regarding the quick reaction force for Yugoslavia, Scharping claimed, "This decision does not come naturally to anyone" (13/48, 30 June 1995: 3959). During that same debate, the chairman of the CDU/CSU parliamentary group, Wolfgang Schäuble, insisted that his party was not agreeing "with jubilance" to the operation and attacked Scharping for having suggested otherwise in the press (Schäuble, 13/48, 30 June 1995: 3965). Through articulations like this, the project of comprehensive security was articulated as compatible (equivalent) with antimilitarism (military reticence) and the lessons of history. In this way, the project reinscribed the historical antagonism into the discourse (if with a changed position of the antagonistic frontier), separating a still inherently peaceful FRG from its aggressive predecessors. The credibility of the project was arguably further strengthened precisely because it was proposed by members of the SPD and Green Party, both of which had been notable for their originally critical stance on military operations (Philippi 1997). The reasoning here, put simply, is that if even former pacifists demand military interventions, those actions surely cannot be entirely wrong.

The project of comprehensive security also incorporated core demands of the project of Germany's international responsibility, most notably the call to accept the country's duty to contribute to peace promotion. In a way, the project of comprehensive security was designed for Germany to assume more responsibility for the promotion of peace and, by extension, to fulfill a moral obligation to those who suffered. By incorporating military operations as an integral part of a comprehensive approach, the project could simultaneously claim to fulfill international expectations. Members of the SPD/Green government referred to the need for Germany to fulfill its "responsibility" (e.g., Fischer, 14/3, 10 November 1998: 112). In that sense, a comprehensive security policy ensured Germany's continued Western and, more generally, international integration. In fact, comprehensive security required multilateral cooperation. Thus, in December 1998, SPD MP Peter Zumkley stressed that "only together with the alliance can we face the secu-

rity political challenges" (14/12, 4 December 1998: 720). A comprehensive
security was inherently multilateral. Through the incorporation of the language of responsibility, the project was able to include demands in the Bundestag to fulfill (ascribed) external demands, which, in various guises (alliance
solidarity and capability, international responsibility, etc.) had been an
important argument during the discursive struggles of the 1990s.

Equally, demands for the respect of international law were incorporated.
This inclusion is most clearly visible in the titles of the motions to grant a
mission mandate, which, without fail, invoke the international legal basis
for the respective operation, in the form of various resolutions by international organizations (most notably by the UN Security Council). For example, the German government's motion to deploy the Bundeswehr in Operation Deny Flight in Bosnia in the early 1990s was somewhat cryptically titled
"German Participation in Measures by NATO and the WEU to Implement
Resolutions by the UN Security Council Concerning the Adriatic Sea
Embargo and the No-fly Zone over Bosnia-Herzegovina" (12/240, 22 July
1994: 21165). By merely listing these resolutions, a relation of equivalence is
constructed (or, rather, implicitly presumed) between the missions put to a
vote and demands for multilateralism, international integration, legitimacy,
and legality (on the link between legality and legitimacy, see Nuñez-Mietz
2018).

Even Operation Allied Force was said to be consistent with international
law, despite a lacking UN mandate and contrary to the opinion of many legal
scholars (see Cassese 1999; Independent International Commission on
Kosovo 2000; less forceful, Simma 1999). Chancellor Schröder insisted that
"it is indisputable within the alliance that international military operations
outside the alliance's area have a clear international legal basis as a precondition." Although he understood the argument by those who saw Allied Force
as illegal, Schröder continued, "After careful consideration I deem it wrong. I
believe that the international legal basis for the NATO operation to contain
a humanitarian catastrophe exists and that it is sufficient" (14/35, 22 April
1999: 2765).

COMPREHENSIVE SECURITY AS A UNIVERSAL REMEDY

As noted in chapter 1, one core ingredient of any successful hegemonic project is the provision of an empty signifier, of a demand that is so broad that it

can function as a symbol for a wide range of demands and, more broadly, the absent fullness of society. In the project of comprehensive security, this function is fulfilled by the demand for such a security strategy. The emergence of comprehensive security did not mean that demands for Germany to accept its new international responsibility vanished from the discourse; that interpretation could not be further from the truth. However, comprehensive security was articulated as the means by which Germany could fulfill that responsibility. One of the strengths of this project was that notions of duty were still very prominent in the discourse, contributing to the overall impression of a process driven by factual necessities rather than choice.

The starting point for the demand for comprehensive security was the perceived inadequateness (a lack) of the established instruments of German security policy to deal, in particular, with conflicts such as the one in Somalia and in (former) Yugoslavia. As Defense Minister Scharping argued in November 1998, Germany and its allies were facing "altered causes of crises, causes of threats, and risks for international security and stability," while "the instruments date from the period after the Second World War, from the period of the Cold War and the block confrontation," and were "not suitable" for the management of the "international crises on our continent and far beyond it." Rather than sticking to the old ways, Scharping argued, German security had to orient itself toward comprehensive security (Scharping, 14/3, 10 November 1998: 114).

Policy on conflict prevention had to adapt in two main ways. First, managing contemporary security problems could no longer be "achieved in the categories of a military alliance alone" (Scharping, 14/3, 10 November 1998: 114) but had to apply a "broad spectrum of political as well as military reaction possibilities and the respective capabilities" (Scharping, 14/35, 22 April 1999: 2771). Second, a modern security policy needed "to identify causes of crises earlier in the future and to act more decisively than in the past" (Scharping, 14/3, 10 November 1998: 113). That need meant that the Cold War reliance on deterrence was obsolete in the face of new threats: "Either we wait until critical developments together with their consequences have arrived here [bei uns], or we face them where they are created" (Scharping, 14/35, 22 April 1999: 2771). The federal government, Scharping argued, embraced the latter option: "crisis prevention has to apply where crises themselves are created" (14/35, 22 April 1999: 2771).

Overall, comprehensive security was articulated as the universal remedy to overcome the new threats in their entirety, a remedy made possible by the combination of different actors' instruments, both civilian and military, into a unified approach. To be more precise, this approach entailed articulating an equivalence between

- civilian and military instruments of foreign and security policy and
- different actors, their instruments, and their demands.

Most important in the context of this study, comprehensive security articulated military and nonmilitary instruments as equivalent instead of disparate or even contrary. During the Cold War, military instruments were articulated as exclusively suited for conventional deterrence and, if entirely unavoidable, territorial defense, while the promotion of peace abroad was entirely conducted by nonmilitary means (e.g., development aid) as part of Germany's military reticence. Within the project of comprehensive security, though, military and nonmilitary instruments of foreign and security policy were articulated as equivalent with respect to the common goal of overcoming the lack of peace.

When the first articulations of the new threats emerged in the early 1990s, those articulations were formulated in explicit opposition to the Kohl government's continued emphasis on military security provision.[7] In his articulation, in 1992, of what he called the "novel dangers," SPD MP Walter Kolbow stressed the new threats as characterized, above all, by not being amenable to military management.

> Facing novel dangers for humanity and the need for assistance *we need an extended, an international understanding of security.* Hunger crises and misery in the countries of the Third World, the enormous problems in the new democracies of Middle and Eastern Europe and in the successor states of the Soviet Union, many developing countries' immense level of debt, environmental catastrophes, refugee flows, and the largely still uncontrolled arms trade entail risks of a global proportion *against which military capabilities are of no use.* Here risks exist, but *they cannot, after all, be removed using military capabilities. . . .* We [as] social democrats have for many years alluded to security problems requiring, *above all, political and economic means.* (Kolbow, 12/70, 16 January 1992: 5882, italics added)

---

7. As Kohl argued, an "anticipatory security policy" had "not become redundant." Due to the emergence of "new challenges and risks" (Kohl, 12/53, 6 November 1991: 4365, 4366), NATO and the Bundeswehr were still needed after the end of the Cold War.

Here, Kolbow not only provided an overview of a range of new threats but explicitly stressed that military means were entirely unsuitable in dealing with them. The new threats would have to be dealt with, "above all," using "political and economic means." In this early articulation, Kolbow stressed the differential content of military means, on one hand, and economic and political means, on the other, as well as the absence of equivalence between the problems to be solved and military means. In short, the logic of difference clearly triumphed here.

A similar articulation of military and civilian instruments prevailed in the original articulations of civilian conflict prevention/resolution, or "civilian crisis prevention" (as it was called by the Schröder government), which would, in a rearticulated version, inform central documents of German policy on conflict prevention, like the 2000 Comprehensive Concept (Federal Government 2000) and the 2004 Action Plan "Civilian Crisis Prevention, Conflict Resolution and Post-Conflict Peace-Building" (Federal Government 2004). In general, civilian conflict prevention refers to the application, as early as possible, of nonmilitary means to prevent or end armed conflicts, through, for instance, diplomatic negotiations, peace pedagogy, development aid, or other measures (Weller 2008; Weller and Kirschner 2005). The concept originally emerged in the context of the German peace movement and peace research. Civilian conflict prevention was initially articulated as an alternative to military conflict management, of which representatives of the peace movement and peace research were highly skeptical. Thus, proponents of civilian conflict prevention like Andreas Buro articulated it as a "political alternative program [Kontrastprogramm]" to the "neo-military-interventionist orientation" of the West (1995: 81). Indeed, civilian conflict prevention was intended as a means to "leave the old military-confrontational paths" (Buro 1995: 81). It was articulated not as equivalent with but as a competing project to military peacekeeping.

At first glance, the project of comprehensive security seems to be highly similar to early articulations of an expanded concept of security à la Kolbow and articulations of civilian conflict prevention. For instance, in April 1999, Chancellor Schröder observed that "after the overcoming of the East-West conflict," security could only "to a decreasing extent be achieved by military means alone" (14/3, 22 April 1999: 5764). However, a second glance reveals significant differences. Where Kolbow stressed the inappropriateness of military means to address the new threats, Schröder only put in doubt that they could be resolved by military means alone. For Kolbow and certainly for the peace movement, military means needed to be replaced by nonmilitary means adequate to address the "novel dangers," whereas Schröder argued for military means to be complemented by other instruments.

This latter understanding of extended or comprehensive security as including military as well as other instruments asserted itself. When the SPD and the Greens won the Bundestag election in 1998 and formed a new coalition government, the demand for civilian conflict prevention became incorporated into the project of comprehensive security, alongside military peace operations; that is, the two approaches were (re)articulated as equivalent demands with respect to overcoming the lack of peace. For instance, in December 1998, Defense Minister Scharping explicitly stressed that one should not understand the general "political line" of crisis prevention with civilian means as an "opposite to the Bundeswehr and to its tasks" (14/12, 4 December 1998: 735). According to the 2004 action plan "Civilian Crisis Prevention, Conflict Resolution and Post-Conflict Peace-Building," civilian crisis prevention "embraces political, diplomatic, economic, humanitarian, and military means" (Federal Government 2004: 10).

Not only were military operations articulated as equivalent with civilian conflict prevention, but they were constructed as the *conditio sine qua non* of civilian efforts for peace. Civilian conflict prevention, it was said, required military operations. This argument had been formulated already in the early 1990s by members of the CDU/CSU/FDP coalition government. For instance, in July 1993, Foreign Minister Kinkel claimed, "Anyone who is not capable to commit, if need be—I explicitly say: if need be—, also militarily, has already in advance, in diplomatic conflict prevention and in other ways as well, a diminished influence" (12/169, 2 July 1993: 14595). This very specific articulation of the relationship between military and civilian means of conflict prevention was adopted by proponents of comprehensive security. For example, in December 1998, Defense Minister Scharping claimed that the Bundeswehr had the "capabilities . . . to make civilian developments in the interest of freedom from violence and peacekeeping possible in the first place" (14/12, 4 December 1998: 735).[8] This incorporation and rearticulation of previously contrary demands made

---

8. For Scharping, this was not a radical shift. Already in July 1994, he had claimed that "questions of prevention" would have to be given priority over "necessary, but not desirable, military means only [to be deployed] as ultima ratio" (12/240, 22 July 1994: 21171).

the project immediately appealing to both advocates of a less militarily oriented security policy and/or of civilian conflict prevention, on one hand, and champions of military peace operations, on the other. Arguably, incorporating demands for civilian conflict prevention makes it more likely that Green MPs would support the overall project, while the incorporation of military operations—as a means of last resort—improved the chance that CDU/CSU and FDP MPs would identify with the project, because it (1) entailed the fulfillment of international demands, particularly those by Germany's European and NATO allies for burden sharing, and (2) picked up claims for the continued relevance and necessity of military security provision—for instance, to exert pressure on "aggressors" to participate in diplomatic negotiations.

Aside from different types of policy instruments, comprehensive security also articulated an equivalence for a number of different actors (subjects) and their demands. Through the articulation of equivalence between various policy fields and instruments, the different actors involved were enlisted for the hegemonic project. More precisely, the project articulated as equivalent different subject positions with which individual subjects could identify (or not). The main point here is that seen from the perspective of comprehensive security, different government ministries and agencies, like the Federal Ministry of Defense (BMVg) or the BMZ, were (claimed to be) united in their common goal to overcome the lack of peace. So were other national governmental actors (e.g., USAID or the French Army), international organizations like NATO or the EU, and nongovernmental organizations active in peace and conflict or development and so on.

In combining all instruments of all actors in basically all fields relevant in the context of conflict resolution, only comprehensive security held the promise of being able to overcome the new threats in their entirety. Neither military means alone nor purely civilian means could overcome all of the new threats. Comprehensive security was the only demand that was contrary to the entire antagonistic chain of equivalences, including armed conflict and other forms of mass violence, underdevelopment, poverty, environmental problems, mass migration, the suffering of refugees, and authoritarian rule. This standing was precisely why comprehensive security could assume the representation of the entire chain of equivalent demands (the Self) and, by extension, the common good as such.

THE FORMATION OF AN ANTAGONISTIC FRONTIER

Through the articulation of equivalence, all of the enlisted demands (e.g., environmental protection) also become articulated as contrary to the new threats as a whole and (if indirectly) to each moment in the totality of the new threats. For instance, by virtue of being equivalent with conflict prevention, environmental protection was articulated as (indirectly) contrary to armed conflict. As a result, the discursive space was split into two opposing (antagonistic) camps. The antagonistic Other was constructed as what I call a "realm of instability" that blocked the world from fully constituting itself as the peaceful and stable place it would otherwise be. This is reflected in, for example, the 1999 claim by former defense minister Rühe that "the enemy of today and tomorrow is called instability" and in the argument by SPD MP Zumkley that "the danger of today is the instability—for different reasons—in some countries and regions" (Rühe, Zumkley, 14/35, 22 April 1999: 2793).

The realm of instability was characterized by authoritarian rule, human rights violations, armed conflict, and insecurity. During the German security debates of the 1990s, dominated by the consecutive violent events in former Yugoslavia and, in particular, the Kosovo conflict, the realm of instability mainly corresponded geographically to Central and Eastern Europe and, specifically, former Yugoslavia. Emerging as a result of this construction of a threatening realm of instability, conversely, was an implicit construction of Western Europe (and "the West" more generally) as a threatened "realm of stability,"[9] marked by democracy, the respect for human rights, peace (understood as the absence of mass violence), and stability.

This split of the discursive realm was most clearly formulated by Foreign Minister Fischer in February 1999, when he argued,

> At this stage, Europe is split in two. When we look at the Balkans, we see the Europe of the past, when we look at Brussels, we see the Europe of integration, the Europe of the future; on one side the Europe of the past, of wars and ethnic cleansing, on the other side the Europe of the future, of integration, and, thank God, of the disappearance of war as a means of politics, the

---

9. The expressions *realm of stability* and *realm of instability* are not taken from the discourse itself. Given the complexity of the discourse and the diachronic and synchronic variation within it, using the two realms as heuristic devices allows for some terminological consistency.

Europe of close cooperation, the overcoming and dissolution of borders. We will have to develop the Southern Balkans toward the Europe of integration. (14/22, 25 February 1999: 1705)

Echoing arguments—about a "modern" West and a premodern, traditional, backward, and threatening non-Western Other—that postcolonial scholars have problematized (Chakrabarty 2000; Mignolo 2007; Quijano 2007), Fischer articulated a split between a "Europe of the future," marked by peace and the disappearance of hypernationalism, and a "Europe of the past," stuck in a violence believed to be long gone from the Continent. While the "Europe of the past" was marked by "wars and ethnic cleansing," the "Europe of the future" represented supranational "integration" (as a departure from past hypernationalist conflicts) and the "disappearance of war as a means" of politics.

Since the threatening outside exported stability, Western Europe, in turn, had to respond through stability export. For instance, Rühe argued that the aim of German security policy must be to "expand the stability achieved in Western Europe to all of Europe" (14/35, 22 April 1999: 2767). One can clearly see here how the realm of instability (the "Europe of the past," or non-Western Europe) was not merely threatening but, at the same time, functioned as a constitutive outside for the realm of stability (the "Europe of the future," or "Western Europe"). Emerging as a result of this exclusion was an image of the Self as a source of stability (see fig. 8 for a stylized overview).

As a "space of stability" in this articulation (Rühe, 14/35, 22 April 1999: 2766), Western Europe was not just threatened but also constituted by the excluded realm of instability. Only through the exclusion of "crises" and "conflicts" as purely a matter of the outside, accompanied by the denial of any continuity between the inside and the outside, could Western Europe / the West constitute itself as a realm of stability. This articulation is, however, just an articulation, not a statement of facts, as illustrated by, for instance, a closer look at the Gulf War, one of the new conflicts. As was also controversially debated in the Bundestag (see, e.g., 11/34, 16 October 1987; 12/2, 14 January 1991), Iraqi president Saddam Hussein received massive support during the Cold War, including arms deliveries by numerous Western governments that arguably enabled him to start a war with Iran in the 1980s and invade Kuwait in 1990 (see Dawisha 1980). Thus, the apparently clear-cut antagonistic frontier between a peaceful, stable, and generally benevolent inside and a threatening outside does not withstand closer scrutiny.

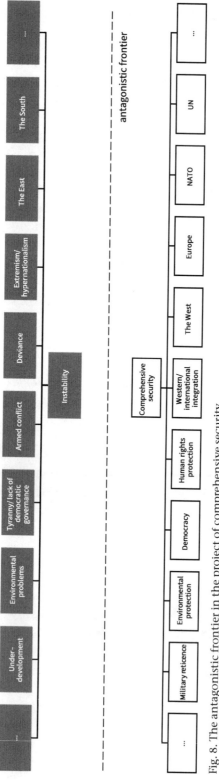

Fig. 8. The antagonistic frontier in the project of comprehensive security

The meaning of the Self-as-realm-of-stability was not precisely tied to Western Europe, as the realm of instability was not necessarily limited to Central and Eastern Europe. For instance, as Schröder pointed out in 1998, problems like ecological destruction were mainly located in "the countries of the South" (14/3, 10 November 1998: 63). Equally, the realm of stability referred to a complex construct of geographical locations and collective subjects, including at least Western Europe and North America, while simultaneously transcending that construct. Decision makers often shifted the subject on whose behalf they spoke (Germany, "Europe" / the EU, "the West" / NATO, the "international community" / UN), often implicitly presuming a general equivalence of these different subjects and their demands in a united quest for world peace. In April 1999, Defense Minister Scharping claimed that NATO was a "central anchor of stability" (14/35, 22 April 1999: 2770), and Rühe, Sharping's CDU/CSU predecessor, stated that "NATO and the European Union have created a space of stability in Western Europe" (14/35, 22 April 1999: 2766). Implied in these blurry, ambiguous constructions is that "we"—namely, Germany, Europe, and the West—speak on behalf not of particular interests but the universal one of the world community as such, which has a legitimizing function (Bliesemann de Guevara and Kühn 2011).

The construction of time is important in this context, as the Other was articulated as located in the past, while the Self was firmly rooted in the present or even in the future. These articulations, in turn, were closely linked to notions of civilization and barbarism. As Foreign Minister Fischer argued in a debate about the budget for the Foreign Office,

> After all, we have made the experience that everything has been tried so that it does not come to the military intervention [in Kosovo], that peacekeeping took priority. That had the consequence that 200,000 people had to pay for it with their lives, that there were mass rapes of women and concentration camps. After all, there was a barbarian displacement policy and, in Bosnia, even a policy of extermination against the Muslims. (14/38, 5 May 1999: 3137)

Through the invocation of concentration camps, Fischer (at least implicitly) constructed a relation of equivalence between, on one hand, the Serbian leadership of the 1990s and, on the other, the Nazi regime, thereby producing a discursive link between the historical antagonistic Other and the Serbian Other.

Such discursive links to Germany's / Europe's own belligerent past were not new at the time but were already present in the articulations of Iraq in the debates about Operation Desert Storm. For example, in January 1991, Chancellor Kohl chastised the Iraqi government for not only the "brutal attack on Kuwait" but also the "barbarian presentation of obviously mistreated allied prisoners of war for the media" (12/5, 30 January 1991: 67). Similarly, in the context of Bosnia, Kinkel argued, in June 1995, that with the Yugoslav wars, "barbarism has literally returned to Europe" (13/48, 30 June 1995). Also, like Fischer, Schröder claimed, in the context of discussions on Allied Force, that NATO could not let "a part of Europe lapse back into oppression and barbarity" (14/35, 22 April 1999: 2762). As Diez has argued (2004: 326), similar articulations equally played an important role in the construction of a (Western) European identity beyond Germany, in which Europe became increasingly articulated in opposition to (among others) Central and Eastern Europe, which were constructed as the "incarnation" of Western Europe's own past.

Constructions of a "barbaric" and / or backward Other draw on long-established discursive patterns that have been present in international politics at least since the era of colonialism, if not since the Crusades (Chakrabarty 2000; Chandra 2013; Grovogui 2010; Muppidi 2012; Salter 2002). As Richard Jackson (2005: 50) has argued with respect to the discourse of the US "war on terror," articulations of "barbaric" attacks on the "civilized world" draw on a discourse about "the noble struggle to civilise the non-western, non-European world," to "bring modernity to the colonies, to save the Somalis and Rwandans from their primitive blood-letting, and to rescue the Kosovars, the Afghans and most recently the Iraqis from their savage rulers." Other examples include arguments, in development discourse, that the West needs to "help" underdeveloped regions (see Ziai 2006, 2014), as well as the recurring statements that some societies are "not ripe for Western democracy" (Risse 2010: 23). Such articulations are intuitively plausible not least because they resonate with established discourses. One could object that colonialism has long since been discredited, particularly its racist aspects, and that drawing a line from colonialism via Rwanda to the present is far-fetched. In this context, however, it does not matter whether colonialism itself has been dismissed as morally reprehensible.[10] What matters is that

---

10. Neither does the argument of the short stint of the German Empire in the colonial business provide grounds for dismissal. National discourses do not take place in the isolation ward, and argumentative patterns, particularly regarding military interventions, tend to be highly similar in Germany and other Western countries.

the discursive patterns that construct a hierarchy between a developed Self and a backward Other, while today usually stripped from obvious colonial and racist undertones, have not vanished but continue to play an important role. Tying articulations to these discursive practices still endows the former with credibility, even if invoking colonialism or openly racist arguments to justify interventions would clearly clash with sedimented practices of a liberal democratic and peaceful Western Self. Moreover, such articulations have political effects, referring non-Western societies back to the "imaginary waiting room of history" (Chakrabarty 2000: 8), marginalizing non-Western voices, and contributing to the legitimacy of patronizing policies (similarly, Bulley 2010; Crawford 2002: 428).

In sum, the project of comprehensive security managed to formulate an overall vision for German policy on conflict prevention, claiming that a comprehensive security approach could overcome the obstacles standing in the way of world peace. This accomplishment also highlights the project's limited scope. The project of comprehensive security did not seek to implement a new grand strategy. Its central concern was not German security but German responsibility to others. Out-of-area operations were primarily justified not on the basis of their contribution to Germany's own security (although vague references to German security are not entirely absent; see, e.g., Federal Ministry of Defence 1994) but by external expectations and moral arguments, regarding Germany's international responsibility. Indeed, in November 1998, Defense Minister Scharping still stated that "also in the future the main task of the Bundeswehr is national defense [*Landesverteidigung*]," instead of out-of-area operations (14/3, 10 November 1998, 115).

Before 9/11, the new threats were articulated more as uncertain risks than as clear-cut security threats (on the distinction, see Kessler 2010). Consider, for instance, the 1996 argument by Green MP Winfried Nachtwei that after the end of the Cold War, "instabilities and risks can be observed that . . . could affect 'national interests' of the Federal Republic and its security" (13/135, 7 November 1996: 12149). Noteworthy here is that Nachtwei's reference to German security remains vague, claiming that it "could" affect Germany's interests or security. He fails to specify exactly how this effect might materialize. His statement is illustrative of the more general articulation of the new threats before 9/11, as something that, if not dealt with, might have some negative effect on German interests and/or security. Indeed, references to "refugee wave[s]," particularly in the context of the Yugoslav wars (see Kolbow, 14/45, 17 June 1999: 3732; Kinkel, 12/101, 22 July 1992: 8610), provide the clearest reference to any potential link between the new threats and German security, and even that securitized representation of mass migration (see, e.g., Diez and Squire 2008) rests largely on metaphor and remains distinctly vague when it comes to any precise explanation of potential negative effects. As a result of this rather tentative, reluctant, partial securitization, the integration of the demand for German security into the chain of equivalences also remained somewhat tentative and partial (even more so than normally), limiting the project's scope.

CONCLUSION

This chapter examined the construction of comprehensive security as a hegemonic project striving to reorganize German policy on conflict prevention. It highlighted, in particular, the construction of various social demands and subjects as equivalent through the exclusion of the new threats as an antagonistic Other as well as through the articulation of comprehensive security as a universal remedy. The result was the emergence of an antagonistic frontier that split the discursive space between a realm of stability and a threatening realm of instability. The analysis here also highlights the important role that constructions of an inferior, backward, non-Western Other played in the production of comprehensive security, demonstrating the continued presence of coloniality in contemporary security discourse. Through the articulation, specifically, of military and civilian means as equivalent in the larger endeavor to overcome armed conflict in particular, advocates of comprehensive security managed to incorporate as equivalent demands that were previously contrary: namely, demands for a stronger focus on nonmilitary instruments, on one hand, and on out-of-area operations, on the other.

However, the project remained limited in scope to German conflict prevention. Comprehensive security was presented not as a blueprint for German grand strategy (i.e., security provision) but, more modestly, as a solution for German conflict prevention, justified mainly by Germany's international responsibility. That presentation changed only after 9/11, when terrorism was incorporated into the new threats and, as a result, the latter's articulation as a radical threat, fundamentally questioning the essence of the Self's identity (and physical security), became much more pronounced. That development is the central focus of the next chapter.

CHAPTER 5

"Forward Defense" as a New Grand Strategy

The Establishment of the New Security Order

The previous two chapters examined how German military operations have been made possible after the end of the Cold War and how the proponents of comprehensive security articulated an overarching vision for German policy on conflict prevention. Despite its display of all the core "ingredients" of an ideal-typical hegemonic project (equivalence, antagonism, and representation), that vision did not gain hegemonic status. Before it could establish itself at the discursive order in the field of peace/conflict prevention policy, 9/11 happened.

The present chapter examines the expansion of the project of comprehensive security after 9/11, its successful establishment of a new security order, and, as a result, the implementation of a new grand strategy that promoted out-of-area operations as a primary means by which German security could be guaranteed under radically changed circumstances. The main argument of this chapter is that the 2001 terrorist attacks did not prove to be a major disruption to the German security discourse, contrary to the US security discourse, which is commonly separated into pre-9/11 and post-9/11 periods (see, e.g., the contributions to Booth and Dunne 2002; Halper and Clarke 2004; Leffler 2003).[1] The German Cold War security order, with its emphasis on traditional military threats and its reliance on military means,

---

1. Both 9/11 and the "war on terror" have been the subject of a number of critical studies (e.g., D. Campbell 2001b; Croft 2006; Edkins 2002; Holland 2009; R. Jackson 2005). Contrary to the bulk of the literature, Nabers (2015) contests the extent to which 9/11 was transformative.

would not have been able to domesticate the 2001 terrorist attacks either, but in 2001, the Cold War order had already been disrupted and had existed in a state of serious disrepair for years. At that point, no overarching alternative had yet been articulated, mainly because discursive struggles of the 1990s concentrated on the more pressing and controversial issue of out-of-area operations. When 9/11 happened, a suitable interpretive framework was readily available, in the German discourse on comprehensive security (contrary to the largely realist US discourse), not only to make sense of the terrorist attacks but to provide a ready-made strategic blueprint for how to deal with such threats. The solution was a comprehensive security policy—a solution that, although primarily articulated in the context of conflict prevention, could easily be adapted to any of the allegedly new, globalized security threats. As a consequence, comprehensive security, or networked security, quickly became established as the dominant framework for post–Cold War German security policy as a whole, providing a grand strategy for the era of asymmetric threats.

This chapter traces that development. Since the (pre-9/11) project of comprehensive security and the (post-9/11) project of networked security overlap to a significant extent, I limit my discussion here largely to modifications of the project, primarily to avoid unnecessary repetition. My general claim is that the project of comprehensive security remained largely intact and was expanded, including a broader equivalential chain that, most notably, now firmly incorporated the demand for security as well,[2] a radicalized articulation of the antagonistic Other, and the construction of the new threats as direct physical security threats instead of uncertain risks. At this point, I will use the term networked security, instead of comprehensive security, for heuristic purposes, to make clear that I am talking about the significantly broadened, post–9/11 version of the original project of comprehensive security.

This chapter is separated into five main sections. The first examines how terrorism has been articulated in the post–9/11 German security discourse, focusing particularly on terrorism's integration into the larger totality of the new threats and, as a result, the latter's securitization. The second section discusses representation, arguing that within the project of networked secu-

---

2. In the following discussion, I use the term security (*Sicherheit*) to refer to the (demand for) the absence of threats to basic values, while the term networked security (*vernetzte Sicherheit*) refers to the approach or policy of combining military and non-military instruments of German foreign policy.

rity (and comprehensive security before it), both the Self and the radical Other became symbolized not by a single signifier but by a set of largely interchangeable signifiers referring to (roughly) the same demand, or what I call a "blurred empty signifier." The third section addresses the (highly ambiguous) articulation of the military as an instrument of German security policy. In the fourth section, I focus more specifically on the politics of networked security, that is, on the many ways in which aspects of the hegemonic project were and remain contested. The chapter's fifth section, an overview of the many ways the project has become sedimented, provides support for my thesis that networked security has become established as Germany's new grand strategy.

MAKING SENSE OF TERRORISM

Two aspects of the articulation of international terrorism after 9/11[3] are of particular importance to this study. The first is that terrorism was articulated in a much more radical fashion than the pre-9/11 new threats. In contrast to the latter, it was articulated not as an ambiguous risk but as a direct physical security threat to Germany, the West, and the world as such; that is, terrorism was securitized. With the terrorist attacks of 9/11, international terrorism emerged as the most important and direct security threat.[4]

To begin with, terrorism was articulated as a physical threat not just to the US but also to German, European, and international security. Shortly after 9/11, members of the SPD/Green government, including Chancellor Schröder, pointed out that 9/11 was "not only an attack on the United States

---

3. In the following discussion, I use the term international terrorism, in line with usage in German parliamentary discourse, although, technically, transnational terrorism is a more precise term, given that it is private (i.e., nonstate) in nature and a cross-border phenomenon rather than one that, technically, is a matter of interstate relations (Baumann and Stengel 2014). For a systematic discussion of different actors in IR, see Genschel and Zangl 2008, 2011. On transnational terrorism and, in particular, al-Qaeda, see Hoffman 2015; Laqueur and Wall 2018.

4. In the years before 9/11, terrorism had only been mentioned sporadically in the German security discourse (e.g., Gehrcke, 14/124, 12 October 2000: 11890). I am not saying that terrorism had not played a role in Germany before 9/11, as anyone familiar with the Rote Armee Fraktion (RAF) knows. However, by the early 1990s, the RAF had all but ceased to exist, and even during its existence, it was primarily a case of domestic terrorism, that is, a law enforcement problem (on the RAF, see Moghadam 2012; Pflieger 2011; Pluchinsky 1993). Only with 9/11 did terrorism emerge as a problem of German external security.

of America" but "a declaration of war against the entire civilized world," including Germany (14/187, 19 September 2001: 18301). That terrorism was articulated as a direct, physical threat to German security is even more clearly visible in other statements. For instance, in regard to a German contribution to the unfolding "war on terror, Schröder argued, "In the decisions that we will have to make, we are guided solely by one goal: to secure the future viability of our country in the midst of a free world; for that is what it is about" (14/187, 19 September 2001: 18302). Similarly, Peter Struck, the chairman of the SPD parliamentary group, pointed out,

> If we agree in the observation that the attacks were directed at the whole civilized world, then, of course, *they have been directed also at us.* . . . *No one should fall into the error [that] the terrorism could roll by Germany* and Europe if we were to keep out of the fight against terrorism now for putative self-interest. (14/187, 19 September 2001: 18307, italics added)

Both of the preceding statements demonstrate that terrorism was articulated as a threat to Germany's existence, to its continued "viability . . . in the midst of a free world," instead of just, say, to its identity, which would be more of an ontological threat. Moreover, the construction of terrorism as a security threat was directly linked to demands (at the time still unspecified) for Germany to become actively involved in the fight against terrorism. Thus, cabinet members claimed that any hope that Germany would remain exempt from terrorist attacks if the country assumed a neutral position was illusory, precisely because terrorism was directed at Germany too. This claim was most clearly expressed by Foreign Minister Fischer on 26 September 2001, when he stated that committing troops to Operation Enduring Freedom was not simply about "abstract alliance solidarity," because Germany would also be "directly confronted" with terrorism "sooner or later" (14/189: 18394). Again drawing on gendered language linking failure to react with strength to passivity and cowardice (Athanassiou 2012; Auchter 2012; Christensen and Ferree 2008; Hooper 2001: 44), Fischer claimed that if the Germans were to "duck away," this would only "inspire" the terrorists (14/189, 26 September 2001: 18395). In that sense, the articulation of terrorism differed significantly from that of the new threats before 9/11, which were articulated as uncertain, ambiguous risks rather than direct physical threats to Germany's security.

The second aspect of particular importance to this study is that terrorism was articulated as a radical Other in the Laclauian sense (i.e., as an ontological threat to the Self's identity), again in a much more aggravated fashion than the pre-9/11 new threats. As noted in chapter 1, while social antagonism and securitization are often closely intertwined, they should not be understood as identical (Rumelili 2015: 57–58), especially in the context of a hegemony analysis, which seeks to explain discursive change. Securitization is crucial for the legitimation of violent policies (Buzan, et al. 1998) and can add to an impression of urgency, but for the formation of a unified subject out of disparate social groups and individuals, the construction of an Other as a common obstacle to the realization of demands (including a fully constituted identity) is much more crucial. Securitization merely identifies that which blocks the demand for security, while antagonism constructs a radical Other that blocks all demands articulated as equivalent.

That situation illustrates how securitization and identity formation are linked from a discourse theoretical point of view (see Stengel 2019b). As the contingent result of hegemony, securitization always involves the (re)production of a referent object, a threatened Self. In that way, identity formation is unavoidably included in securitizing processes. However, contrary to what IR poststructuralism sometimes seems to suggest (e.g., Campbell 1998), identity formation does not necessarily involve securitization. It does necessarily involve antagonism (i.e., the construction of a radical Other as a common obstacle to the realization of "our" demands), but the radical Other does not have to be constructed as a physical threat. This differentiation is of crucial importance in the discussion of the identity-security nexus.

In the case of terrorism, securitization and antagonism coincide. Terrorism was constructed as not just a threat to the world's physical security but a fundamental challenge to the very essence of the so-called "civilized world." In the days after the attacks, Chancellor Schröder claimed that "this kind of violence," which he specified as "the random extinction of innocent human lives," was a challenge to "the basic rules of our civilization," threatening not just "the principles of human coexistence in freedom and security" but "all that which has been built over generations" (14/186, 12 September 2001: 18293). The "faceless and also ahistorical barbaric terrorism," Schröder claimed, was opposed to the world's values.

At the dawn of this century, Germany stands on the right side—one is almost tempted to say: finally-, on the side of inalienable rights of all people. These human rights are the great achievement and the inheritance of European

Enlightenment. These values of human dignity, of liberal democracy, and of tolerance are our great strength in the fight against terrorism. *They are what binds our community of peoples and states, and they are what the terrorists want to destroy. These values, ladies and gentlemen, are our identity, and that is why we will defend them, with vigor, with decisiveness, but also with prudence.* (Schröder, 14/187, 19 September 2001: 18304–5, italics added)

Two aspects of Schröder's statement are especially relevant here. First, he literally stated that terrorism was fundamentally opposed to the Self's very identity. Rather than merely seeking to kill people and destroy things, the terrorists were opposed to the very principles of human coexistence, the core (Enlightenment) values of the civilized world. Second, the radicality of this opposition to the Self's identity is much more pronounced in Schröder's statement on terrorism than in the case of the pre-9/11 new threats. Clearly, the pre-9/11 discourse split the world into two opposing camps, one civilized and the other barbaric (see chapters 3 and 4), but in contrast to tyranny, the terrorist Other is articulated not merely as blocking peace but, being "directed at all which holds our world together at heart," as nothing less than hell-bent on undoing civilization as such. Clearly, terrorism threatened not just people's physical well-being and their property but the Self's very essence.[5] Consequently, the "fight against terrorism" was, at heart, a "defense of our open society . . . , a defense of our liberality and also our way [of] living in an open society" (Schröder, 14/187, 19 September 2001: 18304).

Here, again, one can see the ideological character of antagonism. In the quote excerpted above, Schröder constructed a very specific representation of reality. First, the world that emerges as a result of the exclusion of the terrorist Other is very specific, unified behind the "inheritance of European Enlightenment." Thus, we have here an essentially Eurocentric, or Western-centric, construction, in which the West's values are universalized and proclaimed to be the world's values (Blaney and Inayatullah 2018; Hobson 2012; Matin 2012). Second, Schröder splits the discursive realm into two opposed camps. Although he does not state it in terms as blunt as those used by US president George W. Bush (see Nabers 2015), Schröder constructs a binary choice between either being part of the civilized world or being with the terrorists. Any heterogeneous elements (e.g., groups who neither side with the

---

5. In contrast, as Herschinger (2012: 82–84) has shown, the UN discourse also included more ambiguous and heterogeneous articulations, linking terrorism to liberation movements, for instance.

terrorists nor share Eurocentric Enlightenment values) are denied. Third, by constructing the world as split between a peaceful, liberal democratic "civilized world" and a "barbaric," "inhuman" terrorism threatening it with violence from the outside (Fischer, 14/189, 26 September 2001: 18395), a mythical, benign image of the Self is constructed, in which negative developments like violence, a lack of democracy, or human rights abuses are articulated as firmly located on the outside. What emerges is the image (articulation) of a peaceful and democratic world that has been attacked for no reason. As Zehfuss has aptly put it, in such articulations, "war arrives from what is imagined as the outside" (2018: 4). As noted in chapter 1, this highly selective reading of the situation denies the multiple ways in which Western policy itself is implicated in the reproduction of violence, inequality, and injustice. It also locates responsibility for violence squarely with the Other. As Schröder claimed in October 2001, "We—that applies to us all—did not want this conflict. It has been forced upon us by barbarian attacks in the United States" (14/192, 11 October 2001: 18680).

Although a detailed discussion is beyond the scope of this chapter, it is important to note that this articulation of terrorism as a radical Other was not limited to Schröder or to members of his government but extended across party lines as well as time. For instance, during the debate on 19 September 2001, then CDU chairwoman Angela Merkel stated, "We will have to draw the lines anew. . . . They will be drawn between democracy and dictatorship, between respect for human rights and disregard for them, between freedom and oppression" (14/187: 18326).[6] Equally, this articulation was not

---

6. Further examples abound. Friedrich Merz, chairman of the CDU/CSU parliamentary group, argued that the attacks were attacks "on the civilization, on the freedom and the openness of our societies" (14/186, 12 September 2001: 18294). Consequently, the terrorists were "enemies of the open society" who "challenge the basic values of democratic and liberal societies," which was why "freedom has to be defended anew now" (Merz, 14/187, 19 September 2001: 18305, 18306). Defense Minister Scharping (SPD) called 9/11 an "attack on our values" (14/187, 19 September 2001: 18327); his predecessor, Rühe (CDU/CSU), labeled it an "attack on us all" (14/189, 26 September 2001: 18396); deputy chairman of the CDU/CSU parliamentary group Michael Glos called it an "attack on our way of life" and called on Germany to "defend civilization and democracy" (14/187, 19 September 2001: 18318, 18319). Quite similarly, the chairmen of the Green and the FDP parliamentary groups—Rezzo Schlauch and Wolfgang Gerhard, respectively—interpreted 9/11 as an attack "on our civilization" (Gerhard, 14/186, 12 September 2001: 18295) or "the open society in general" (Schlauch, 14/186, 12 September 2001: 18296). Equally, the chairman of the PDS parliamentary group, Roland Claus, articulated it as an "attack on the civilian society, on culture and humanity" (14/186, 12 September 2001: 18296).

limited to the immediate aftermath of 9/11 but extended in time. For example, in a debate about the Kunduz air strikes in September 2009, the CDU/CSU parliamentary group's speaker for foreign affairs, Eckart von Klaeden, claimed, "These terrorists hate us not for what we do but for what we are. That is why we may not give in here" (16/233, 8 September 2009: 16308).

Although the construction of the terrorist Other seems to be a textbook case of social antagonism, it would be misleading to argue that terrorism functioned as a radical other for the overall security discourse, because that would conflate different levels of discourse. Instead, I argue that the statements quoted above are part of an emerging counterterrorism discourse within the wider security discourse. They refer mainly to terrorism as a distinct phenomenon, paving the way primarily for German participation in the "war on terror." It would be misleading to see these statements as an indicator for the articulation of antagonism at the level of the overall security discourse, because they are much more limited in scope. This limitation becomes immediately transparent if one takes a closer look at what is or is not justified with a reference to the terrorist threat. Although individual operations, most notably Operation Enduring Freedom in Afghanistan (OEF-A), were legitimized with the terrorist threat, one is hard-pressed to find arguments referencing terrorism alone to justify the continued maintenance of the Bundeswehr or force transformation. Thus, terrorism did not provide an overall rationality for German security and defense policy as such, like the Soviet Union did during the Cold War. Much more relevant for this particular study is how terrorism was integrated into the wider security discourse, which is focused not just on one (if, admittedly, the most important) threat but on the security situation as a whole, on the basis of which grand strategy was formulated. This brings us to the second important aspect of the articulation of terrorism: its integration into the new threats.

Crucially, terrorism was articulated not as an isolated phenomenon but as an integral part of the new threats. Because of the availability of comprehensive security as an interpretive framework, 9/11 could immediately be made sense of as an example of exactly the kind of new threats already spoken about by proponents of comprehensive security. Thus, while stressing the magnitude of the attacks, German policy makers pointed out that the phenomenon of international terrorism as such was neither entirely new nor encountered out of the blue. This was clearly put by Defense Minister Scharping on 19 September 2001.

In light of this threat, *which is not new but whose quality, extent, and effectiveness have now become horribly visible*, what NATO's heads of state and government have already formulated in [NATO's] 1999 [Strategic Concept] will maybe be more understandable, namely, *that crisis prevention, comprehensive security policy, and, included in it, the fight against international terror are common tasks*. (Scharping, 14/187, 19 September 2001: 18324, italics added)

As that statement from Scharping most aptly illustrates, not only was terrorism not a radically new threat (see also Beer, 14/204, 28 November 2001: 20129), but it only demonstrated what advocates of comprehensive security had known all along: namely, that comprehensive security, including counterterrorism, was the appropriate response to today's security environment. This view was shared by the CDU/CSU as well. Even more to the point, Friedrich Merz, chairman of the CDU/CSU parliamentary group, claimed that 9/11 was "the first test case for the new NATO which, already with the strategic concept of 1999, has adjusted to the changed security situation" (14/187, 19 September 2001: 18305), and CDU chairwoman Merkel pointed out that "the threats of the 21st century" had "on 11 September, *at the latest*, gained a clear face. . . . No one can say anymore that he has not seen it. *All warnings about such dangers have been outmatched by reality*" (14/187, 19 September 2001: 18326, italics added).[7] As these statements illustrate, German decision makers claimed that terrorism was not new but merely a confirmation of the existence of what Merkel called the "threats of the 21st century." Moreover, while early statements refer to the terrorist attacks as a declaration of war, the overall interpretation of 9/11 as acts of war that need to be responded to in kind—prevalent in the US (R. Jackson 2005: 38–40)—did not take hold in Germany, where counterterrorism was explicitly articulated as a case for a comprehensive, or networked, approach.

Because 9/11 could be rendered intelligible within the discourse of comprehensive/networked security, the latter was not disrupted but, instead, sig-

---

7. Compare, however, the interpretation of the situation by former defense minister Volker Rühe on 26 September 2001: "The series of barbaric terror attacks in the US confronts us with a fundamentally new situation. We all agree on that. What hitherto has, under the heading 'asymmetric threat,' been abstract theory has become gruesome reality these [past few] days" (14/189: 18396). Here, Rühe points out that the situation is "fundamentally new." At the same time, however, he already offers a framework within which to interpret the events, that of asymmetric threats, which, Rühe himself points out, have been around as "abstract theory" prior to 9/11.

nificantly broadened, incorporating terrorism as an integral part of the new threats. Thus, political actors constructed causal links between new phenomena like terrorism and piracy and the previous elements of the new threats. They argued that comprehensive security was also the right answer to terrorism because the latter, similar to armed conflict or other "crises" (to use the term most prominent in the German discourse), was closely linked to contributing factors or root causes—indeed, the very same causes in which the "crises" of the 1990s were rooted. This articulation of terrorism and other phenomena as equivalent began with references to the Taliban as an example of authoritarian rule. Already one day after the terrorist attacks, Schröder stated that "anyone who helps terrorists or protects them offends all fundamental values of the coexistence of the peoples" (14/186, 12 September 2001: 18294), and when Afghanistan began to materialize as the first target in the "war on terror," articulations increasingly focused on the Taliban (e.g., Struck, 14/210, 22 December 2001: 20832).

More generally, from the beginning, terrorism was discussed in conjunction with what decision makers referred to as terrorism's "breeding grounds" (Nährboden) (Schröder, 14/186, 12 September 2001: 18294). Building on the discursive connections already established between armed conflict, various forms of extremism, and other "causes," political actors quickly made connections between terrorism and various contributing factors, already familiar from prior articulations of the "causes" of crises (see chapter 4). Discourse participants articulated a relation of equivalence between terrorism and "the causes and breeding grounds of terror: conflicts, poverty, ignorance and disease" (Zapf, 14/189, 26 September 2001: 18399). To overcome terrorism, they argued, armed conflict, poverty, a lack of education, and disease needed to be addressed as well. Here, again, one can see how the construction of equivalence and antagonism are joined at the hip.

As was the case during the 1990s, different discourse participants articulated different but partially overlapping sets of new threats, producing a broad antagonistic chain of equivalences on the level of the overall discourse (see also Schlichte 2008). Struck, chairman of the SPD parliamentary group, identified "poverty, social misery, and hurt pride" as part of terrorism's "societal sounding board" (i.e., its breeding grounds) (14/187, 19 September 2001: 18308), SPD MP Detlef Dzembritzki pointed to "poverty, environmental destruction, hunger, and violence" as the causes of "refugee movements and migration" and "international terror" (14/189, 26 September 2001: 18423), and Barbara Hendricks (SPD), parliamentary state secretary in the Ministry of Finance, claimed that "poverty and a lack of participation often are the breeding grounds for fanaticism and [they] characterize nondemocratic regimes" (14/199, 9 November 2001: 19531). Equally, organized crime (in particular, money laundering) and financial deregulation were said to contribute to terrorism (Schröder, 14/187, 19 September 2001: 18303; 14/192, 11 October 2001: 18682), and, in November 2006, the CDU/CSU parliamentary group's speaker for foreign affairs, Eckart von Klaeden (16/64, 10 November 2006: 6319) claimed that "the question of the proliferation of weapons of mass destruction" could not be "separated from Islamist extremism, from failing states anymore," linking both WMD proliferation and state failure to terrorism.[8]

In November 2003, Andreas Schockenhoff, the CDU/CSU parliamentary group's deputy speaker for foreign affairs, argued, with respect to Congo and other "failed" or "failing" states,

> *States in decay or on the verge of decay in Africa and elsewhere become ideal suppliers for organized crime and international terrorism.* Consequently, it is a big task of development policy to disarm these time bombs for the long term and to prevent the emergence of even more terrorism and crime. (15/73, 7 November 2003: 6297, italics added)[9]

Failing or failed states that could no longer exert control over their territories were considered potential safe harbors for terrorists and criminals. This articulation was reproduced over time, with German decision makers continuously arguing that state failure in Afghanistan had made 9/11 possible (e.g., Merkel, 16/214, 26 March 2009: 23122). As a consequence of the threat posed by state failure, networked security had to involve state- and nation-building.[10]

---

8. I here understand state failure broadly, as states' loss of their ability to exert control over their territory (see Rotberg 2002). In the literature, this loss is also called "state collapse," "state decay," "fragile statehood," or "limited statehood" (see Doornbos 2002; Krasner and Risse 2014; Risse 2005; Schlichte 1998; Schneckener 2007). The descriptor "state failure" is not neutral but rests on a Eurocentric understanding of the state that has political consequences (Figueroa Helland and Borg 2014). It is also questionable to what extent the state in its current OECD form ever was a dominant feature of world politics (Schlichte and Wilke 2000).

9. On a different occasion, in the debate about EUFOR RD Congo on 19 May 2006, Schockenhoff argued that "also the Congo exports organized crime and streams of refugees to Europe" (16/36: 3106).

10. While state-building aims at the sustainable "strengthening [of] state structures

Illustrative here is a statement from April 2005 by Hans Martin Bury (SPD), minister of state in the chancellery.

> In the periphery of Europe and in farther-off regions, the increasing decay of state structures leads to armed groups and nonstate actors gaining ever more influence. The consequences are terrorism, organized crime, corruption, as well as human and drug trafficking.
>
> In our globalized world, these are not regionally limited phenomena anymore; in many ways, they also endanger the security of the international community. (Bury, 15/172, 21 April 2005: 16084)

In addition to the articulation of equivalence between state failure, armed conflict, terrorism, organized crime, corruption, human trafficking, and drug trade, particularly remarkable here is the geographical expansion entailed in the shift from comprehensive to networked security. While policymakers during the late 1990s stressed the importance of developments "on our doorstep" (Schröder, 14/3, 10 November 1998: 63), Bury claimed that even developments in "farther-off regions" cannot be ignored anymore. As a consequence, networked security required an interventionist, preventive strategy with a scope that, in principle, considered the whole world. Seen here is how the discourse draws on the globalization discourse for credibility, which stresses increasing "societal denationalization," including the growth of denationalized/globalized policy problems that affect societies across great distances but, at the same time, remain beyond the problem-solving capacity of individual states (Zürn 2013: 403; critical, Coward 2018).[11]

The result of the overall articulation is an emerging antagonistic chain of equivalences that poses a common obstacle to the realization of a terrorism-free, perfectly secure world:

> ... ≙ terrorism ≙ armed conflict ≙ poverty ≙ ignorance ≙ fanaticism ≙ tyranny ≙ lack of participation ≙ hurt pride ≙ social misery ≙ environmental destruction ≙ refugee movements ≙ migration ≙ disease ≙ organized crime ≙ human trafficking ≙ drug trade ≙ corruption ≙ uncontrolled financial flows ≙ WMD proliferation ≙ state failure ≙ ...

---

and institutions," nation-building focuses on the overall "development of the entire society," including the promotion of a national identity (Schneckener 2007: 10).

11. See also, e.g., Castells 2011; Genschel and Zangl 2013; Leibfried et al. 2015.

As a consequence of this broad construction of the new threats (of factors responsible for the continued lack of peace and security), a wide chain of equivalent demands also emerged on the flip side of the antagonistic frontier. Like in the 1990s, conflict prevention, poverty reduction, democratization, and so on were articulated as equivalent. New demands like counterterrorism, the fight against organized crime, preventing WMD proliferation, the fight against diseases, corruption, human trafficking, and so on were equally incorporated into the chain, significantly increasing its breadth and breaching super-differential boundaries that separated previously distinct discourses (e.g., the health discourse or discourses on financial regulation) from the security discourse.[12]

In this context, it is important to note the important role that terrorism plays in the construction of the overall discourse, which functions as an anchor point in the discourse. Through the articulation of terrorism as an integral part of the new threats, the latter were rearticulated as (contributing to) physical security threats; that is, by virtue of their tight connection to terrorism, the new threats as a whole were much more closely linked to (in) security than they had been before. As noted in chapter 4, in the 1990s, security was only indirectly and tentatively linked to the project of comprehensive security. After 9/11, terrorism symbolized a clear lack of security, which, in turn, made it possible to incorporate the demand for security into the overall project. In that sense, (counter)terrorism assumes the role of a privileged signifier, making possible the incorporation of the demand for security in the project of networked security. At the same time, it does not assume the role of an empty signifier. This situation illustrates that the structure of hegemonic projects is much more complicated in terms of stratification than empirical studies often seem to suggest, displaying a hierarchy including the empty signifier among a much wider range of more or less important moments in the discourse.

At the same time, the incorporation of security meant the redefinition of national defense, rearticulating interventions as a defense measure. This rearticulation was most clearly put by Defense Minister Scharping in September 2001, when he argued that "national defense under the changed security political circumstances means the acquisition or the development of capabilities that can also be applied in crisis prevention as well as in crisis reaction" (14/189, 26 September 2001: 18408). Scharping's statement dem-

---

12. On the securitization of health, see Hanrieder and Kreuder-Sonnen 2014.

onstrates that while defense remained the main task for the Bundeswehr, what defense meant in practice changed significantly, from conventional deterrence to crisis prevention and reaction. This shift is equally visible in statements by Scharping's successors Peter Struck (SPD) and Franz Josef Jung (CDU/CSU). For instance, Defense Minister Struck claimed, in a debate about the "new Bundeswehr" (as he called it),

> The protection of Germany remains a core task of the Bundeswehr. It even has gained a new, more comprehensive meaning; for in addition to national defense in the traditional sense, [which has] become less likely, the protection of our population and vital infrastructure from terrorist and asymmetrical threats has to be ensured. (15/97, 11 March 2004: 8601; see also Jung, 16/60, 26 October 2006: 5784)

This incorporation of security is crucial, for in the context of security discourses, "security" is not just a demand like any other. Instead, "security" refers to the "specific universal" (Nonhoff 2017a: 91, 2019: 76) of the security discourse. Security (a general absence of threats) is what the security discourse is all about.[13] Consequently, being able to incorporate the demand for security into the chain of equivalence makes it possible for a specific project to compete for the position of security order, to attempt, at least, to achieve hegemony in the overall security discourse. Put simply, only because of terrorism could the project of networked security hegemonize the security discourse.

Beyond the demand for security, the project of networked security incorporated a number of demands that had not been part of the project of comprehensive security, including immigration control, energy security and cybersecurity, international trade, economic policy, and so on.[14] Some of

---

13. That the name of the discourse and the specific universal (the common good in the context of a specific discourse) share a signifier—somewhat confusing in this case—is coincidental and does not have to be so. For instance, in the social policy discourse, the specific universal would be welfare or prosperity or something similar. This does not have to be a single signifier; indeed, since the universal is unattainable, it is not really directly representable in discourse. Since fullness cannot ever be reached (no meaning can ever be fully fixed), the specific universal is an empty place rather than a clearly identifiable "thing." As the specific universal here, "security" refers to a complete absence of any threats to virtually any value—a state that realistically can never be achieved.

14. For instance, on 29 October 2002, Schröder connected counterterrorism to economic demands, arguing that the "dangers of international terrorism" and the

these demands—for example, those for secure energy supplies, cybersecurity, or free trade—link to previously unconnected discourses beyond the security discourse, which significantly strengthens the project's chance of success. Moreover, these more specific demands are linked to very broad demands like justice, tying in with more general discourses about basic values beyond even foreign policy. As Fischer pointed out on 26 September 2001, finding a lasting solution to terrorism required the creation of "an [international] order . . . that is based on human rights, democracy, justice, and sustainability and that strives to induce a balance of interests in the hot conflicts of this world" (14/189: 18395).[15] The overall goal was nothing less than "to integrate, if possible, all countries into a global system of security and prosperity" (Schröder, 14/187, 19 September 2001: 18302). Thus, inherently linked to the construction of the new threats is a much broader vision (an imaginary) of a perfectly secure, peaceful, free, just, and united world society, to be realized only if the new threats could be overcome. Because only networked security could vanquish the new threats, it also, symbolically at least, represented a fully constituted, perfect, peaceful, and secure world community.

In addition to articulating a number of demands as equivalent, the project of networked security attempted to construct a unified collective subject—namely, a (unified) world. When comprehensive security was formulated as a strategy for German conflict prevention policy, the need to coordinate the activities of different state and nonstate actors was deemed a crucial element. Foreign policy is usually aimed at influencing developments outside a state's own territory (Carlsnaes 2002; Cohen and Harris 1975; Hill 2016), so even the most powerful actors usually have less than complete control over outcomes (on foreign policy implementation, see Brighi and Hill 2008).[16] Moreover, foreign policy unavoidably entails the presence of "rival" actors, whose activities do not necessarily have to be compatible with states' foreign policy goals (Baumann and Stengel 2014; Chong 2002; Stengel and Baumann 2018). Advocates of networked security (and comprehensive security before that) expected that a "modern" security policy necessarily had to

"dangers of regional conflicts . . . threaten our domestic security but also our economic prosperity" (15/4: 51).

15. See also Schröder, 15/4, 29 October 2002: 58; Scharping, 14/189, 26 September 2001: 18408.

16. I am leaving aside the question of to what extent states, even in the OECD, can control what goes on inside their territories (see, e.g., Genschel and Zangl 2017; Leibfried et al. 2015).

be a coordinated endeavor, involving different state and nonstate actors. As Defense Minister Scharping concisely put it in the context of counterterrorism, "The communality of the response [to terrorism] has as much to do with its comprehensive character as the comprehensive character [has to do] with the communal response" (14/189, 26 September 2001: 18408).

Beyond these rather practical questions, the project of networked security also articulated a number of subjects as equivalent, either presuming or declaring equivalence for their demands (i.e., their interests). Most basically, the project claimed such equivalence for the interests of Germany, Europe, North America, and even the entire (civilized) world. This understanding is clearly visible in the numerous quotes above that construct the new threats, particularly terrorism, as a security threat not just to the US but to the "civilized" world and the international community as such. It is presumed that the world is united in a common interest to combat terrorism. For instance, on 12 September 2001, Chancellor Schröder stated,

In reality—this becomes increasingly clear—we already are one world. That is why the attacks in New York, the place of business of the United Nations, and in Washington are directed against us all. *Yesterday's terrorist attack has shown us quite plainly: In our world, security cannot be separated. It can only be achieved when we stand even closer together for our values and work together in their implementation.* (14/186: 18293–94, italics added)[17]

The project of networked security proclaimed that the security of Western states and their populations were equivalent also with the security and well-being of people living in those regions in which the new threats (allegedly) originated, that is, the populations to be subjected to intervention. Thus, national (German, US, etc.), international, and human security were claimed to go hand in hand. According to the project, only humanity as a whole or no one could be fully safe. Thus, the antagonistic frontier was clearly located between the world (and humanity as basically undivided), on one hand, and terrorists, criminals, pirates, dictators, and other shady figures, on the other. This understanding manifests itself as well in the more

---

17. The discourse is ripe with statements that proclaim the convergence of interests between Germany, its allies, and various international organizations, like, for example, the 2011 claim by Defense Minister Thomas de Maizière that "also alliance interests are most times at the same time our national security interests" (17/112, 27 May 2011: 12816).

specific debates about individual military operations, which were articulated as actually being in the interests of the very civilian population they often endangered.[18] Even those military operations conducted mainly for the sake of German/Western security (those not primarily motivated by "humanitarian" concerns) were argued to be in the interests of the local population.

Among many examples is a statement about OEF-A that Chancellor Schröder made in November 2001.

> Anyone who has seen the television images of the celebrating people in Kabul after the withdrawal of the Taliban—above all, I think of the images of women who can finally meet freely in the streets again—,
>
> (Applause by the SPD, the Green Party and the FDP as well as representatives of the CDU/CSU and the PDS)
>
> should not have difficulties evaluating the result of the military strikes [as being] in the interest of the people there. (Schröder, 14/202, 16 November 2001: 19856)

While primarily justified with Germany's own security interests, OEF-A was also said to help the Afghan people, by which claim the latter were incorporated into the chain of equivalences. This argument was repeated in time and across party lines. For example, in the first debate focused on the 2009 Kunduz air strike on 8 September 2009, Chancellor Merkel claimed that ISAF

> contributes to the protection of international security, worldwide peace, and life and limb of the people here in Germany from the evil of international terrorism. That stood at the beginning of this mission, and that applies until today. That is our conviction. This found and finds the agreement of the Afghan government, and we know how many ordinary Afghans again and again ask us not to leave them alone in the fight against the Taliban. (16/233: 26298)

Both these quotes aptly illustrate that "the West" and the Afghan population are articulated as equivalent. This articulation applied more generally to various others, most notably the civilian populations of countries to be

---

18. This argument has been discussed at length in the critical literature on (supposedly ethical) Western wars, "humanitarian" intervention, and the responsibility to protect (e.g., Bellamy and Dunne 2016; Bulley 2010; Chandler 2015; Orford 1999; Sabaratnam 2018; Zehfuss 2018).

subjected to intervention. Because Western interventions (broadly under-stood) guaranteed both Western security and the peace, security, and free-dom of, for example, Afghans, no tough decisions were necessary between what could otherwise be disparate or even contradictory interests.

Thus, more broadly, the project of networked security articulated self-interest and altruism as compatible (see also Ziai 2010 in the context of development policy): both ensuring German security and promoting peace, security, and freedom for others required a networked (interventionist) secu-rity policy. This compatibility was most aptly summarized by Defense Minis-ter Jung in October 2006, who claimed that "crisis and conflict manage-ment . . . corresponds to our values, our mission, and our interests" (16/60, 26 October 2006: 5784). This claim that no tough choices needed to be made between "our" interests and the interests of others is a core strength of the project. At the same time, it is worth pointing out that this articulation, too, is ideological. The still unclear future of Afghanistan (e.g., Congressional Research Service 2019; International Institute of Strategic Studies 2019; Mur-tazashvili 2016) supplies but one example showing that the equivalence between "our" and "their" interests is highly debatable.[19]

That example also helps illustrate how articulations appear convincing by drawing on other established discourses. Here, again, established gen-dered and racialized discursive patterns particularly lend credibility. Already, the routine references to the "civilized world" point to a problematic and highly ideological construction of Self and Other, in which violence and other problems are externalized (see Salter 2002). Gender researchers have argued that femininity and victimhood are so bound up that they have started referring to "women and children" with one word, as "womenand-children" (Carver 2006: 453; Enloe 2014: 1). We see similar arguments in vir-tually all German debates about military operations, in which the local pop-ulation, particularly the population of women and children, is articulated as in need of Western help.[20] As a result of sedimented discursive patterns that articulate the non-Western Other as backward, German policymakers claim

---

19. Almost two decades since the beginning of the US-led intervention, its benefit for either Western or Afghans' security is yet to materialize. Not only does the country remain mired in armed conflict, which has caused more than 10,000 civilian casual-ties (killed or injured) annually over the past years (International Institute of Strategic Studies 2018; Shortland et al. 2019; UNAMA 2018), but al-Qaeda, the original reason behind the intervention, seems far from defeated (Celso 2018; Lefèvre 2018).

20. See, e.g., Dunn 2008; Grovogui 2010; Hudson 2012; Inayatullah 2014; Mup-pidi 2012; Shepherd 2006; Stern 2011; Welland 2015.

to know, as a matter of course, the most pressing security concerns of "local populations" in the non-West as well as the most appropriate problem solutions, even though the "locals" never get to contribute to the conversation (Bertrand 2018).

Moreover, like the Cold War security discourse before (see chapter 3), the discourse of networked security is ripe with references that express the need to be grateful to the soldiers, which, by invoking Young's (2003) logic of masculinist protection, works to delegitimize (ungrateful) critique of the armed forces and, by extension, security policy. For instance, in a debate about ISAF on 28 September 2005, Defense Minister Struck claimed that "on their mission, our soldiers are exposed to dangers to life and limb" and that they deserve a "high [degree of] respect for their work and all our gratitude" (15/187: 17573). This claim was usually tied to the demand for broad support for the operation. Thus, in November 2004, Struck argued, "Our soldiers have a claim [haben einen Anspruch darauf] that the parliament support [trägt] this operation with a broad majority" (15/139, 12 November 2004: 12784). Here, Struck claims that the soldiers and their (self-proclaimed) advocate, the government, have a right to demand that the Bundestag support the operation.

One additional strength of the project of networked security was its flexibility, its ability to domesticate new demands and events. Over time, the project incorporated newly emerging policy problems into the new threats, simultaneously broadening the chain of equivalent demands. Most notable among the new threats added over the years were piracy, cybercrime, cyberwarfare, and cyberterrorism, as well as threat to the German energy supply. When piracy off the coast of Somalia (re)emerged as a security problem in the middle to late 2000s, as sporadic incidents turned into more organized forms and as raids of trade vessels significantly increased in number (Anning and Smith 2012: 28), piracy was also integrated into the construction of the new threats. Specifically, piracy off the Somali coast became connected to state failure, organized crime, and terrorism. As, for example, Foreign Minister Steinmeier (SPD) put it on 17 December 2008, piracy was said to threaten "the last remnants of order . . . on which the people in Somalia depend," the payment of ransoms would "further strengthen the criminal structures and undermine the Somali state even further," and since piracy was causally linked to organized crime and state failure, these problems could only be addressed together: "Only . . . if state structures in Somalia are restored will it be accomplished to truly end piracy" (16/195: 21057–58). Other discourse participants linked piracy to terrorism. For example, FDP MP Rainer Stinner argued that terrorism could "often not be separated from organized crime and piracy" and was using sea trade for the "illegal transport of weapons and persons" (16/185, 4 November 2008, 19756).[21] Finally, cybercrime, cyberwarfare, cyberterrorism, and the disruption of energy supplies were equally incorporated into the totality of the new threats. For instance, in November 2010, Schockenhoff argued,

> International terrorism and organized crime as well as instability that is based on failing states threaten the entire civilized world. . . . The proliferation of weapons of mass destruction and missiles has immediate consequences for our security. The consequences of climate change can lead to conflicts for natural resources or settlement areas and to large migration streams with security political repercussions for us. Cyberattacks and potential attacks on trade routes and our energy supply are new dimensions of the concrete threats to our country. (17/71, 11 November 2010: 7602)

These examples clearly show that the category of the new threats is highly flexible and allows discourse participants to link their pet issues to networked security. This flexibility is one important element of the project's overall success. Because the new threats are a highly fluid category (an empty signifier), it is relatively easy to link specific phenomena to it and articulate them as new or asymmetric threats. This also illustrates how the construction of security threats (securitization) can function in at least two ways (Stengel 2019b). First, new threat constructions can emerge as part of the implementation of a new security discourse (as the new threats were securitized after 9/11). Here, acceptance of threat constructions primarily rests on whether that specific order manages to become hegemonic—that is, on antagonism, equivalence, representation, and credibility. Second, issues previously unconnected to security can be incorporated into an existing security order, as happened with piracy and cyberterrorism in the German security discourse. In that case, what matters most is whether new articulations

---

21. See also Homburger, 16/197, 19 December 2008: 21344; Polenz, 16/197, 19 December 2008: 21353; Merkel, 17/37, 22 April 2010: 3477. This view was not uncontested, however. For instance, SPD MP Kurt Bodewig stated, "Pirates . . . are no terrorists" (16/197, 19 December 2008: 21345). Equally, Defense Minister Jung rejected the reinterpretation of Operation Enduring Freedom to also include counter-piracy, due to constitutional constraints (16/185, 4 November 2008: 19757).

("securitizing moves," in Copenhagen parlance) are compatible with accepted articulations, that is, if a credible argument can be made that they are a threat. Here, compatibility is not a matter of a phenomenon's objective essence but is, instead, the contingent product of discursive struggles (Nonhoff and Stengel 2014).

NETWORKED SECURITY AS A BLURRED EMPTY SIGNIFIER

One additional important feature of any successful project is the provision of a symbol with which subjects can identify. Remarkable in the case of the project of networked security is that this role was fulfilled not by a single signifier (narrowly understood) but by a cluster of similar signifiers that referred to roughly the same demand, what I have called a "blurred empty signifier." So far, I have referred to the radical Other mainly as the new threats, primarily for consistency. However, different discourse participants used a wide range of terms, also changing over time, to refer to the set of new problems that characterized the post–Cold War security environment as opposed to the old Soviet threat.

Already, the project of comprehensive security was marked by overwording in regard to the radical Other. References included

- "new hazards" for "peace and freedom" (Kohl, 12/5, 30 January 1991: 67),
- "new challenges and risks" (Kohl, 12/53, 6 November 1991: 4366),
- "novel dangers" (Kolbow, 12/70, 16 January 1992: 5882),
- "new insecurities and dangers" (Schäuble, 12/70, 16 January 1992: 5898),
- "entirely new security political challenges" (Kinkel, 12/151, 21 April 1993: 12925),
- "new international challenges" (Spranger, 13/31, 30 March 1995: 2429),
- "new challenges" (Rühe, 13/31, 30 March 1995: 2415; Schröder, 14/35, 22 April 1999: 2764),
- more "diverse security political challenges" (Kohl, 13/65, 27 October 1995: 5567),
- "instabilities and risks" (Nachtwei, 13/135, 7 November 1996: 12149),
- "new instabilities" (Schröder, 14/3, 10 November 1998: 63),
- "challenges to our security" (Scharping, 14/35, 22 April 1999: 2772),

- "security political challenges" (Struck, 15/97, 11 March 2004: 8602),
- "instability" (Zumkley, 14/35, 22 April 1999: 2793), or
- "new risks" (Polenz, 14/124, 12 October 2000: 11877).

After 9/11, the range of terms expanded even further, adding

- "new international challenges" (Merkel, 14/192, 11 October 2001: 18686),
- "the threats of the 21st century" (Merkel, 14/187, 19 September 2001:18326),
- "new threats to our security" (Polenz, 15/172, 21 April 2005: 16086) or simply "new threats" (Meckel, 15/172, 21 April 2005: 16074),
- "entirely new, often asymmetric threats" (Bury, 15/172, 21 April 2005: 16084),
- "risks and threats" (Kolbow, 15/105, 29 April 2004: 9506), and
- "new dangers" (Schäuble, 15/10, 14 November 2002: 540).

The post–9/11 increase in references to "threats," instead of "risks" and "challenges," reflects the increased securitization of these phenomena.

All of the aforementioned terms refer to the novelty of the new threats as well as to the fact that we are dealing with a set of interrelated problems, while different discourse participants listed different phenomena (see the discussion in chapter 4). This flexibility of the discourse across different phenomena is relevant, I argue, for the appeal of the overall hegemonic project, because using different terms allows discourse participants both to appropriate the concept as their own and to stress the problems that they see as most important. The vagueness of the category of the new threats, its flexibility, makes it possible for a wide range of discourse participants across party lines to identify the new threats as the core problem and also to give the problem their specific "spin" and make it their own.

A similar proliferation of signifiers took place with respect to the empty signifier symbolizing the Self's demands and identity as well as the common good more generally, that is, "networked security." While "networked security" functioned in principle as an empty signifier, this side of the antagonistic frontier is equally marked by overwording, as the terms comprehensive security and networked security already illustrate. In addition to these two terms, a number of other signifiers were equally used to refer broadly to the unified, preventive approach to security policy (combining the civilian and military means of different actors) for which discourse participants were calling. Since Kolbow's early (and, to my knowledge, earliest) argument for an "extended, . . . international" concept of security policy (12/70, 16 January 1992: 5882), a number of different terms have proliferated referring broadly to the kind of strategy he had envisioned. Already in July 1994, Defense Minister Rühe (CDU/CSU) and the SPD parliamentary group's speaker for disarmament, Uta Zapf, fought over who had "invented" the concept of "comprehensive security," the Kohl government or the opposition, with Rühe arguing that, "of course," the "political stabilization and the economic stabilization" of the "new democracies" in Eastern Europe had priority "also for security political reasons" (Zapf, Rühe, 12/240, 22 July 1994: 21185), thus calling for a whole-of-government strategy. As discussed in chapter 4, under Chancellor Schröder, who took office in 1998, "a comprehensive security policy" emerged as the shorthand for the government's policy of combining military with civilian means in a whole-of-government approach (see, e.g., Defense Minister Scharping, 14/3, 10 November 1998: 113). This is what others called a "political-military overall approach" (Rühe, 14/35, 22 April 1999: 2766), an "extended security concept" (Beer, 14/189, 26 September 2001: 18413; Wieczorek-Zeul, 14/55, 16 September 1999: 4928), or an "extended security precaution" (Krönig, 14/75, 1 December 1999: 6904).

After the change of government in September 2005, in which Merkel (CDU/CSU) assumed the leadership of a CDU/CSU/SPD grand coalition, "connected security" (*verbundene Sicherheit*) and, in particular, "networked security" (*vernetzte Sicherheit*) were introduced as new terms referring, again, roughly to the same broad demand, for a unified approach instead of a purely military one. Networked security emerged as the core term describing the Merkel government's approach to security policy. Like comprehensive security, networked security, which was introduced by Defense Minister Jung and presented the core concept of the 2006 white paper on defense (Federal Ministry of Defence 2006),[22] referred to the integration of "military, but also developmental, economic, humanitarian, police, and intelligence instruments of conflict prevention and crisis management" (16/60, 26 October 2006: 5785). Indeed, during the debate on the white paper, Jung (16/60, 26 October 2006: 5785) himself used the terms comprehensive security approach, connected security, and networked security interchangeably.

---

22. In quite patriotic language, Jung argued that networked security "runs like a black-red-golden thread through the white paper" (16/6026 October 2006: 5785).

The 2006 white paper established networked security as Germany's new grand strategy (Jung, 16/227, 18 June 2009: 25170). Moreover, the term networked security (or, alternatively, networked approach) was adopted by Chancellor Merkel (16/214, 26 March 2009: 23121; 17/37, 22 April 2010: 3479) as well as Jung's successors. Both Theodor zu Guttenberg (17/15, 20 January 2010: 1315), who succeeded Jung as defense minister after the 2009 Bundestag elections and the formation of a CDU/CSU/FDP government, and Thomas de Maizière (17/112, 27 May 2011: 12816; 17/237, 25 April 2013: 29784), who replaced zu Guttenberg after he (very reluctantly) resigned over his plagiarized doctoral dissertation (Jansen and Maier 2012), adopted the term. So did other government officials, like Foreign Minister Guido Westerwelle (FDP) (16/214, 26 March 2009: 23126) or Christian Schmidt (CDU/CSU), parliamentary state secretary in the Ministry of Defense (17/11, 16 December 2009: 829). Indeed, comprehensive security and networked security were often used interchangeably. For instance, in January 2010, zu Guttenberg argued in favor of a "comprehensive, indeed networked approach of all actors and their means" (17/15, 20 January 2010: 1315). In that sense, the project illustrates that, contrary to a recent argument by Rothe (2015: 45), securitization does not require that "security" functions as an empty signifier under all circumstances. Seemingly necessary are the incorporation of security as a demand in the chain of equivalences and the articulation of the antagonistic Other as contrary to the demand for the security of the respective referent object (the subject or object supposedly threatened by it).

The introduction of new terms to refer to comprehensive security did not mean that the old ones vanished from the discourse. Instead, a proliferation of more or less synonymous terms are used interchangeably. For instance, in October 2006, Kolbow, the deputy speaker of the SPD parliamentary group and former minister of state in the BMVg, referred to the approach as based on an "extended security concept" but also referred to the overall approach as "networked security policy" (16/60, 26 October 2006: 5788). Thus, the function of representation in the German security discourse is assumed not by a single term but by a conglomerate of terms that roughly refer to the same demand: namely, some form of combination of civilian and military instruments of different government agencies and actors. The empty signifier is, in that sense, blurred.

Like in the case of the radical Other, the blurriness of the empty signifier representing the common good contributes to the appeal of the overall project. For instance, the Merkel government's introduction of networked security as an alternative to comprehensive security, which had been the Schröder government's core concept, allowed the former to put its own stamp on what was, at least broadly, the same approach and to claim it for itself, which makes sense in terms of party competition (see Lehmbruch 1998: 19–24). Also, the (partial) rebranding of comprehensive security lowered the threshold to identify with the project for subjects that otherwise might have remained skeptical—most notably advocates of a strong defense, or of primarily military operations for the purpose of alliance solidarity, and those skeptical of civilian conflict prevention—without repelling its previous advocates.[23]

Thus, overwording facilitates the spread of the project across party lines. A similar diffusion can be observed between different ministries. During the Schröder government, comprehensive security was mainly introduced as part of a larger initiative to spread civilian crisis prevention (Federal Government 2004) that was led by the Foreign Office. Networked security was introduced as the central concept in the 2006 white paper, the core strategy document of the Ministry of Defense. The introduction of networked security made possible the embrace and appropriation of comprehensive security across party lines and departmental boundaries. The term networked security also connects the project more closely to notions of globalization that emphasize a supposedly more "networked" contemporary world (Castells 2011). Already, the articulation of the new challenges draws heavily on the globalization discourse, arguing that the challenges have to be addressed at the point of their origin because of their globalized nature, but the term networked security makes immediate sense in the context of globalization, which is often discussed in relying on the metaphor of a network (Coward 2018). This relationship becomes even clearer if one uses a different English term for the German *vernetzte Sicherheit*, which can also be translated as "interconnected security." Interconnectedness is one of the central topics in the omnipresent debates about globalization (see, e.g., Castells 2011; Zürn 2013), which makes the notion of interconnected or networked security

---

23. To criticism of the 2006 white paper by the Greens, Hans-Peter Bartels (SPD) responded that networked security was a "clear affirmation of a comprehensive security concept, not limited to the military, that runs like a red, green, and now also black thread [referring to the colors of the SPD, the Greens, and the CDU] through the whole text [of the white paper]" (16/60, 26 October 2006: 5794). Also, other members of the SPD (Annen, 16/145, 21 February 2008: 15357) and the Greens (Nachtwei, 16/189, 28 November 2008: 20410) adopted networked security as basically a new instantiation of comprehensive security.

immediately appealing as a concept.[24] In an interconnected world, it seems self-evident that security policy would have to be interconnected as well, an understanding that further strengthens the credibility of the overall project.

In a way, the term networked security is not much more than an incomplete case of rebranding (because comprehensive security continued to be used synonymously), providing a surface of inscription for a broad range of demands. Put simply, the vagueness of the concept allows discourse participants to associate their specific demands and priorities with the concept. For instance, networked security can refer to a demand for increased German participation in military operations, as long as they include some form of civilian activities as well, even if they are subordinated to a military command. However, it could also refer to a demand for the role of the Bundeswehr to be limited only to an auxiliary function, if any, within a mainly civilian approach and under civilian command.[25] Not surprisingly, some political actors emphasized the Bundeswehr as an indispensable ingredient of networked security, while others stressed the priority of civilian instruments.

The term networked security also provides an example for how over-wording can be a double-edged sword, enabling a wide group of subjects to identify with a project while simultaneously opening up a space for new discursive struggles to emerge, as the convergence of different signifiers is contested. Because an empty signifier is always only tendentially empty, discursive struggles emerge over whether different signifiers really do refer to broadly the same demand. The blurred signifier itself can become the subject of the logic of difference. This happened in the context of networked security, because the shift from comprehensive to networked security entails a move from the primarily civilian approach of the Foreign Office, subsumed under the headings of "comprehensive security" and "civilian crisis prevention" (Federal Government 2004), to the primarily military approach of the Ministry of Defense (Federal Ministry of Defence 2006). While both concepts stress the importance of integrating both military and civilian initiatives (of both state and nonstate actors), they differ with respect to who takes

---

24. I thank Martin Nonhoff for pointing this out.

25. For instance, in the debate about the 2006 white paper, the chairwoman of the Green parliamentary group, Renate Künast, pointed out, with respect to terrorism, that the Greens "do not deny that the fight against international terrorism has to be conducted also with military means; but we know after all that civilian means must always have intellectual priority" (16/60, 26 October 2006: 5793).

the lead, a difference that provided a source of constant debate about the precise meaning of networked/comprehensive security. For instance, in a debate about Afghanistan, Gernot Erler, deputy chairman of the SPD parliamentary group, claimed that the CDU/CSU/FDP government's "doctrine of so-called networked security" was met with "broad disapproval" by development agencies and nongovernmental organizations, which felt "under pressure . . . to put themselves into the service of military goals" (Erler, 17/85, 21 January 2011: 9557). That example demonstrates how the shift from comprehensive to networked security increases the number of subjects identifying with the project but, at the same time, fuels tension.

THE AMBIGUOUS ARTICULATION OF MILITARY OPERATIONS

Military operations loom large in any debate about security or conflict prevention. As Brockmeier and Rotmann observe with respect to the prevention of atrocities, "To this day, every debate . . . gets short-circuited into a debate about military intervention" (2018: 21). The privileged role of military operations in parliamentary discourse is at least partly due to the legal requirements of the Bundeswehr as a "parliamentary army" (see, e.g., Gareis 2010; Meiers 2011; Meyer 2006; Wiefelspütz 2009). Put simply, any military mission has to be debated in and approved by the Bundestag, while civilian policy measures fall squarely within the executive's foreign policy prerogative. As a consequence of this sedimented discursive practice (established by the BVerfG's 1994 out-of-area decision), military operations will unavoidably receive more attention in the Bundestag, simply because the executive cannot act without parliamentary approval.

The prominence of military missions in parliamentary debates also has to do with their normative contestedness (in light of antimilitarism), which brings us to the discussion of precisely how they were articulated. As noted in chapter 4, the demand for military operations was included in the overall hegemonic project, articulating that demand as equivalent with policy goals such as (German, Western, international, and human) security and peace. For example, in 2001, Defense Minister Scharping pointed out that "crisis prevention is no contradiction to [military] crisis reaction, [just] as crisis reaction, conversely, must never be made into a replacement for or contradiction to necessary prevention" (14/189, 26 September 2001: 18408). Military operations were rearticulated as compatible with—even the precondition for the deployment of—other policy instruments, such as development aid. This relationship is most clearly visible in the construction of OEF-A. For instance, in December 2001, Chancellor Schröder argued,

> It counts among the bitter truths that the peace in Afghanistan has only come closer because of war. It counts among the lessons of the recent German history, which we all have experienced together, that pseudoreligiously legitimized and motivated violence had to be abrogated and overcome by democratically legitimized counterviolence. (14/210, 22 December 2001: 20822)

Like networked security in general, military operations were said to be for the benefit of those subject to intervention. In the context of OEF-A, for example, Schröder argued, "War strikes innocents. That is no question. But the example [of] Afghanistan shows: Only with the aid of military violence could [it] be prevented that also in the future innocents have to suffer indefinitely" (14/210, 22 December 2001: 20822). Similarly, in another example for the important influence of gender discourse, Foreign Minister Fischer claimed that "severe human rights violations and, above all, the oppression of the rights of women and girls" in Afghanistan would have continued without the intervention (14/210, 22 December 2001: 20826). This claim articulated the application of military violence, which at least entails the acceptance of potential civilian casualties, as equivalent with sedimented articulations of the Self as inherently peaceful and benevolent, an articulation making the maintenance of a radical discontinuity with the (violent) past possible.[26]

Also in the project of networked security, military means were articulated as indispensable. As Scharping argued, for example, terrorism could not be subdued by focusing—"almost like a social worker" (*in fast sozialarbeiterischer Weise*)—exclusively on prevention, which did "not . . . suffice" against those already committed to violence (14/189, 26 September 2001: 18408). Similarly, in the debate about the 2006 white paper, Defense Minister Jung claimed that "an actively formative security policy requires a capable Bundeswehr" (16/60, 26 October 2006: 5785), and Kolbow, deputy chairman of the SPD parliamentary group, called the Bundeswehr "an indispensable part of a networked security policy" (16/60, 26 October 2006: 5788).

---

26. See Zehfuss 2018 for extensive discussion of the complicated role of ethics in the legitimation of war.

Analogous to arguments in the later 1990s, military means were articulated after 9/11 as a precondition to the application of civilian instruments. Thus, ISAF was said to be a necessary requirement to "consolidate the commenced stabilization process" (Struck, 14/210, 22 December 2001: 20832), creating "the parameters for the economic development and the societal democratization" of Afghanistan (Struck, 15/17, 20 December 2002: 1313). In the debates on specific operations, Germany's civilian efforts even became an argument invoked as a reason for MPs to vote for the prolongation of the military mission. For example, in 2003, Struck justified his demand for a prolongation of the (military) ISAF mandate by arguing,

> In the next months, we will see whether the experiment, started by Germany, to place *civilian reconstruction* in the foreground of the activities in Afghanistan will succeed or not. It *will succeed so much the better if the German Bundestag provides the soldiers with as broad a vote of support as possible.* That is what I sincerely ask of you. (15/70, 24 October 2003: 5990, italics added)

If MPs support the primacy of civilian reconstruction, the argument went, they should vote for the military mandate.

At the same time, the construction of military means in the project of networked security was highly ambiguous. Thus, military operations (or any application of military means, including the threat of force) were articulated as a necessary instrument but, at the same time, as a means of last resort. Given that I have already discussed these matters in chapter 3, two examples suffice here. In his speech calling for German participation in OEF-A, Chancellor Schröder claimed, "No one takes the decisions [necessary for the deployment of troops] lightly—neither do I. But they are necessary, and therefore they have to be made" (14/202, 16 November 2001: 19857). Similarly, in a debate about the EU's mission in the Congo in April 2010, Schröder's successor, Merkel, claimed that "military restraint and the application of military means as [an] ultima ratio" was part of the "raison d'état of the Federal Republic of Germany" (17/37, 22 April 2010: 3477). These quotes reveal the instability of the relation of equivalence between different moments of the discourse of networked security—military operations, peace, antimilitarism, and so on.

These connection between these discursive moments is unstable for at least two reasons. First, deeply sedimented antimilitarist practices in German security discourse make it difficult to permanently link military means to the demand for peace. After all, for years, the use of force had been understood as fundamentally opposed to achieving peace as well as morally reprehensible. Second, this articulation became even more complicated within the broader articulation of the new threats. In particular, epistemic sedimented practices (i.e., what is commonly held to be factually true) stood in the way of constructing a stable articulation of equivalence between military means and the goal of overcoming the new threats. While there is some acknowledgment that military operations can contribute to the solution of some policy problems, most notably for traditional peacekeeping (e.g., Salvatore and Ruggeri 2017), it is widely acknowledged that they are entirely unsuitable solutions for other challenges, including most of the new threats—to name but a few, climate change, state failure, organized crime, poverty, or corruption. These issues are generally not amenable to a military solution. As a consequence, linking military means to the new threats as a whole can only be credibly done by way of what could be called "discursive hedging," that is, by linking any demand for military operations to a simultaneous acknowledgment of their limited utility and moral undesirability.

This example of instability also shows how discourses do not "float freely," to appropriate Risse's formulation (Risse-Kappen 1994). While the openness of discourses means that different discourses can influence each other, that openness does not mean that one discourse simply gets adopted in a different context. Indeed, the German case demonstrates that discourses need to be rearticulated (translated: see Stritzel 2014) to make them credible against the sedimented practices relevant in that particular context. Although a detailed discussion is beyond the scope of this chapter, numerous discursive elements of the "war on terror"—the presumption that terrorism could be fought primarily with military means, for instance, or the articulation of equivalence between Iraqi president Saddam Hussein and al-Qaeda (see Eberle 2019; Nonhoff and Stengel 2014)—were not credible for a German audience and had to be rearticulated. Thus, the counterterrorism discourse in Germany is significantly different from the US discourse of the "war on terror."

More generally, maintaining the precarious link requires continuous discursive work. Part of this work is constantly stressing that refusing military operations equals inactivity, which commonly meets disapproval due to established gendered behavioral standards (Carver 2008; Hooper 2001; Sjoberg 2010). Moreover, it equally requires various practices of discursive delineation (logic of difference), separating out-of-area operations from interventionism and war, both of which were and continue to be closely associated with Germany's imperialist, aggressive, and authoritarian past (see chapter 3). Thus, advocates of networked security stressed that force transformation was "not about establishing an intervention army and meddling, as some wrongly fear, without need in the affairs of other states but about advocating [our] common security together with our allies and partners, where it is necessary" (Struck, 15/97, 11 March 2004: 8603; equally, zu Guttenberg, 17/15, 20 January 2010: 1316).

The preceding quote equally points to the close interconnectedness of national, international, and transnational discourses and to how speakers in the German Bundestag draw on other discourses and statements by various external authorities for intertextual legitimacy. For example, discourse participants regularly cite leaders of international organizations (IOs), such as the UN or NATO secretary-general (e.g., Jung, 16/60, 26 October 2006: 5783). Discourse participants often attempt to claim that their position has the support of prominent leaders (see, e.g., Lafontaine, Nachtwei, Wiesskirchen, 16/2, 8 November 2005: 48, 49, 51). In the German discourse, intertextual references to IOs or allied states carry a lot of weight because they invoke sedimented practices of multilateralism and Western integration. For instance, citing the UN or NATO secretary-general not only utilizes the latter's expertise but also implies that disagreement equals a deviation from the UN or NATO position, which involves a conflict with multilateralism and/or Western integration.

Even more important in this context is the firm rejection of any connection between Bundeswehr operations and war (see, e.g., Geis 2019; Martinsen 2013; Shim and Stengel 2017), which often manifested itself in a form of "verbal gymnastics" (Ray 2013: 97). For example, in November 2001, Angelika Beer (Greens) claimed, "Our soldiers do not consider themselves soldiers in uniform with a weapon in their hand" (14/204, 28 November 2001: 20130). In 2004, Struck even claimed that the Bundeswehr was "Germany's biggest peace movement" (15/97, 11 March 2004: 8604).

More specifically, with respect to Afghanistan, Struck clarified in 2005, "We do not conduct a war mission but a peace mission" (Struck, 15/187, 28 September 2005: 17574). Even when this articulation became increasingly difficult to reconcile with reports about a deteriorating security situation in

Afghanistan, particularly in the wake of the 2009 Kunduz air strike, discourse participants remained extremely reluctant to refer to the situation in Afghanistan as a war. When Defense Minister zu Guttenberg expressed sympathy for German soldiers calling the Afghanistan mission a war, citing what he, quite cautiously, called a "warlike situation," he immediately went on to argue that, from a legal perspective, that mission could not be called "a classical war. International law is clear here: Wars can only be waged between states, which was why the situation in Afghanistan should be described in terms of international law as a 'noninternational armed conflict'" (17/3, 10 November 2009: 85).[27] Zu Guttenberg repeated this statement a few days later in an interview with the German tabloid *Bild* (zu Guttenberg 2009).[28]

Already, this reluctant phraseology on zu Guttenberg's part was far from uncontroversial. Thus, SPD MP Hans-Peter Bartels pointed out that, "of course," the situation in Afghanistan was "no war," and he commended zu Guttenberg for clarifying this point after the aforementioned *Bild* interview (Bartels, 17/15, 20 January 2010: 1317). Similarly, on 27 January 2010, SPD chairman and former federal minister Sigmar Gabriel called for German decision makers to "stop dealing with the term 'war,' or 'belligerent conflict,' as frivolously as during the last months," claiming that the UN did "not wage a war" in Afghanistan and that "our soldiers are no warriors there" (17/18, 27 January 2010: 1525). Such articulations reinscribe into the discourse the historical antagonism according to which war is something conducted by the FRG's predecessors and its opponents but not by the FRG itself. In my opinion, this reinscription contributed significantly to the credibility of demands for out-of-area operations, because calling Germany's own actions "war" would very likely clash with sedimented practices of military reticence and articulations of an inherently peaceful Self.

---

27. In a similar fashion, others also only spoke of "warlike conditions" (Beck, 17/11, 16 December 2009: 849; Merkel, 17/18, 27 January 2010: 1524; Schmidt, 17/7, 26 November 2009: 393; zu Guttenberg, 17/7, 26 November 2009: 388) or "belligerent situations" (Willsch, 17/15, 20 January 2010: 1331).

28. In contrast, in a 2009 interview with *Internationale Politik*, the Green parliamentary group's speaker for security policy, Winfried Nachtwei, stated, in no ambiguous terms, "It is . . . appropriate to say that there is a war going on in the Kunduz region and that the Bundeswehr participates in that war" (Nachtwei 2009: 2).

CONTESTATION AND THE FAILURE OF ALTERNATIVE ARTICULATIONS

The project of networked security was not uncontested either, and neither was its predecessor. In fact, SPD and Green MPs during the 1990s and Linke MPs ever since then have challenged various aspects of what would become the project of networked security. A detailed and exhaustive discussion of that challenge is beyond the scope of this chapter, but let us consider it in relation to three particularly important aspects: (1) the articulation of equivalence between military operations, on one hand, and peace and security, on the other; (2) the construction of the new threats as a common Other; and (3) the articulation of equivalence between military operations and civilian instruments of foreign and security policy.

First, contrary to what the discussion so far might have suggested, the articulation of equivalence between out-of-area operations and policy goals like peace or security have been subject to contestation since the early 1990s. During the 1990s, members of both the SPD and the Greens voiced opposition to out-of-area operations. For example, already in 1992, SPD federal manager Günter Verheugen expressed the fear that German participation in any operations to intervene in the Yugoslav wars might escalate violence instead of containing it (12/150, 26 March 1993, 12872–73). Green MP Hans-Christian Ströbele, by harshly criticizing the government in 1999 and saying that he was "ashamed for his country that now *again* wages war in Kosovo and that *again* throws bombs on Belgrade" (14/30, 25 March 1999: 2423, italics added), explicitly challenged the imaginary of a new, inherently peaceful Germany. The most sustained critique came from members of Die Linke in the context of the "war on terror," who claimed that war was exactly what Germany was doing in Afghanistan[29] and who criticized the German out-of-area operations as an expression of "militarization."[30] In September 2009,

---

29. For instance, in January 2008, chairman of the Linke parliamentary group Gysi criticized the government for having "the illusion of being able to fight terrorism with war" (16/139, 24 January 2008: 14640). Similarly, Jan van Aken, member of the foreign affairs committee and deputy chairman of the Linke parliamentary group, charged the government with hiding that "in your war innocent civilians die" (17/9, 3 December 2009: 672); two weeks later, he claimed that members of the CDU/CSU "speak about well construction and withhold the corpses" (17/11, 16 December 2009: 839; see also Schäfer, 17/15, 20 January 2010: 1320).

30. This criticism has been expressed multiple times over the years (see, e.g., Lederer, 12/70, 16 January 1992: 5887; van Aken, 17/3, 10 November 2009: 77; Höger, 17/74, 24 November 2010: 8128; Hänsel, 17/191, 12 September 2012: 22966).

the chairman of the Linke parliamentary group, Oskar Lafontaine, put the humanitarian aspects of ISAF in doubt, arguing that "with a lot less money [than the operation cost], one could protect distinctly more people from the death by hunger and from the death by disease without having to kill even a single other person" (16/233, 8 September 2009: 26305).

In particular, the Afghanistan operation was highly controversial (Lagassé and Mello 2018), and critics voiced doubts that military operations were a suitable means to fight terrorism. For instance, speaking for opponents of the intervention in the SPD, MP Rüdiger Veit argued in November 2001, "We do not consider the deployment of the Bundeswehr in Afghanistan . . . right; for in our opinion war is no suitable means in the fight against terrorism" (14/202, 16 November 2001: 19891).[31] Indeed, its fiercest critics even claimed that the ISAF mission decreased the security of Germany. Lafontaine, for example, succinctly claimed, in a debate about the Kunduz air strike, "We cause the opposite of what we actually want to cause" (16/233, 8 September 2009: 26305). Green MP Ströbele argued that the government's increasingly offensive strategy in Afghanistan would contribute to the war becoming ever "worse and more ruthless" and ultimately promoting terrorism (17/7, 26 November 2009: 400).

Second, critics equally challenged the articulation of equivalence between different elements of the new threats. To give but one example, in debates about the counterpiracy Operation Atalanta off the Somali coast (see Peters et al. 2014), Linke MPs questioned the connection between terrorism and piracy, emphasizing difference over equivalence. MP Norman Paech, for example, criticized that in the context of piracy, al Qaeda was "brought into play to connect the fight against terror with the fight against the pirates. This mixing [of other missions] with the war of OEF [the counterterrorism operation] we already know from Afghanistan. To that the Linke says: not with us" (16/195, 17 December 2008: 21063).

Third, the articulation of equivalence between military and civilian means has been subject to sustained criticism, most notably by members of Die Linke. For instance, in September 2001, Wolfgang Gehrcke, deputy chairman of the Linke parliamentary group, argued that the

---

31. Nevertheless, Veit pointed out that if faced with the vote of confidence, he and other critical SPD MPs would vote in favor of the motion—a decision interpreted by Andrea Nahles, chairwoman of the SPD's youth organization Jusos, as an expression of "a pacifist position that retains its capacity for politics [Politikfähigkeit]" (14/202, 16 November 2001: 19884).

Schröder government had "so far conducted a policy that consisted of spending less for development and diplomacy and more for the military." Speaking for the Linke, he continued, "We want the exact opposite: more for development and for diplomacy and less for the military" (14/189, 26 September 2001: 18404). Here, in an example for the logic of difference, Gehrcke challenges the claim that the interests of different ministries were aligned in a comprehensive/networked approach, and he points instead to competition over scarce resources (also Schäfer, 16/227, 18 June 2009: 25171).

More generally, Kathrin Vogler, Linke member of the subcommittee on civilian crisis prevention and networked security, criticized the Greens for their articulation of civilian and military forms of conflict prevention as equivalent, which meant the "subordination of the civilian to military structures." She explained, "Civilian conflict management for you [the Greens] is precisely not the *alternative to* military violence, but it is supposed to *complement it* within the framework of comprehensive, interagency, civil-military concepts" (17/121, 8 July 2011: 14342, italics added).[32] For Vogler, civilian conflict prevention and comprehensive security were contradictory rather than equivalent. Others criticized the "blending of the civilian and the military," arguing that it was "a contradiction in itself to claim that military instruments could contribute to civilian crisis prevention" (Hänsel, 16/74, 15 December 2006: 7466).

In sum, numerous aspects of the project of networked security were challenged. Why the challenges fell on deaf ears has to do, I argue, with mainly two factors: (1) the content of the articulations and (2) the speaker position of their advocates. First, as is already visible in the reactions to the discursive interventions of the Greens in the late 1980s (see chapter 3),[33] many of the articulations challenging the project of networked security (and comprehensive security before it) clashed with the basic normative foundations of

---

32. Similarly, in a debate about the German command of the Quick Reaction Force (QRF) in Afghanistan, Gysi emphasized the differential content of military force and civilian reconstruction, arguing that since a big part of the QRF was the targeted killing of enemy forces, it "had nothing to do with educating girls" (16/139, 24 January 2008: 14640), that is, the civilian goals cited by other MPs.

33. The articulation of the Green Party as somehow dubious and/or untrustworthy by more established parties in the Bundestag had remarkable staying power. Still in 2000, 17 years after the Greens had entered the Bundestag for the first time and also after they supported Operation Allied Force in 1999, FDP MP (and former paratrooper) Dirk Niebel asked whether the Greens' former demand for the abolition of the Bundeswehr still applied (14/124, 12 October 2000: 11886).

German parliamentary discourse. In particular, Linke MPs (like the Greens in the 1980s) questioned the historical antagonism according to which the (inherently benign and peaceful) FRG had nothing in common with its aggressive predecessors. For instance, in a debate about NATO's new Strategic Concept in November 2010, Gesine Lötzsch, deputy chairwoman of Die Linke's parliamentary group, called NATO "this organization that bears [the] responsibility for the deaths of countless civilians and [that] has made the lives of millions of people more precarious [*unsicherer*]" (17/071, 11 November 2010: 7605). Such articulations conflict with decades of discursive practices inscribing and reinscribing into the discourse an inherently peaceful articulation of not just the FRG but the West in general. Similarly, in November 2008, Linke MP Lafontaine argued that insofar as military strikes violated the legal requirement "to spare innocents, . . . such operations to combat terrorism themselves become acts of terror" (16/2, 8 November 2008: 48). Such statements that link the Self to terrorism equally clash with deeply engrained understandings of the Self as benign.

Not surprisingly, policy demands by Die Linke were regularly dismissed on practical grounds as unrealistic and populist, even by Green MPs. For instance, in a debate about the UN Mission in Sudan (UNMIS), the Green speaker for defense Winfried Nachtwei raked Die Linke over the coals for their proposal to send civilian personnel instead of soldiers. According to Nachtwei, the proposal was "so absurd, so preposterous [*abenteuerlich*] and irresponsible that it should be made known to the public" (16/33, 7 April 2006: 2774). In general, the party was dismissed as irresponsible, populist, and motivated by party political concerns rather than by a serious contemplation of the issues (also zu Guttenberg, 16/49, 19 September 2006: 4816).

Second, in the case of Die Linke (and, equally, in the case of the Greens at least until the early 2000s), lack of credibility has been further aggravated by the way the party has been constructed in the Bundestag. Mainly due to the party being the SED's successor, Die Linke is commonly held to be untrustworthy and lacking credibility particularly on matters of democracy and peace. To give an example, in the debate about the first ISAF mandate, Struck, chairman of the SPD parliamentary group, denied the PDS (as Die Linke was then still called) the right to advocate for peace.

One word to the PDS: A party that has succeeded the SED and consequently . . . also has to accept responsibility for [*muss stehen zu*] those who prosecuted peaceful citizens of the then GDR because they wore patches [with the inscription] "Swords into Ploughshares" in no sense has the right to act as [a] peace party in this Bundestag. (Struck, 14/210, 22 December 2001: 10831)

Because its predecessor had been responsible for the prosecution of peace activists in the GDR, Die Linke could not be taken seriously as an advocate of peace today.

Here, the reason for rejection is not so much the content of the statements but the subject position of PDS/Linke MPs. Given that the GDR had itself been articulated as part of the tyrannical Other (see chapter 3), advocacy by PDS and Linke MPs for peace and democracy had a particularly ironic ring to members of the established parties. Not surprisingly, statements dismissing arguments by PDS and Linke MPs are widespread in German parliamentary discourse. For instance, Michael Glos (CDU/CSU) called the party the "SED that now calls itself PDS" in 1998 (13/248, 16 October 1998: 23150), and Beer (Greens) reprimanded the PDS in 1999 for what she saw as "pure populism: The 'peace party' PDS! Embarrassing, embarrassing!" (14/75, 1 December 1999: 6898). Since, to the more established parties, the GDR itself represented an instance of authoritarian aggression (much like the Soviet Union and the FRG's own past), the successor to its leading party, the SED, was simply untrustworthy.[34]

THE INSTITUTIONALIZATION OF THE NEW SECURITY ORDER

The project of networked security displayed all the characteristics of a successful hegemonic project. It managed to establish a new security order and, consequently, to transform German grand strategy. In this section, I briefly discuss the discourse's hegemonization and institutionalization, focusing on four aspects: (1) statements by discourse participants, (2) the codification of networked security and out-of-area operations in policy documents, (3) changes in legislation, and (4) institutional reform and innovation.

First, discourse participants increasingly began to incorporate policy successes into the chain of equivalences, referring to both networked security and out-of-area operations. For instance, in the context of the 2009 Kunduz

---

34. See, e.g., Dressel, 16/120, 24 October 2007: 12518; contributions to the debate in 16/139, 24 January 2008: 14631-47.

air strike, chancellor Merkel claimed, "The policy of networked security today [is] consensus among the allies. That is a sustainable success of German Afghanistan policy" (16/233, 8 September 2009: 26299). Discourse participants connected successful past operations to demands for future operations (Schockenhoff, 16/49, 19 September 2006:4814) and argued that these missions had contributed to the Bundeswehr's "standing . . . in the community of peoples" (Struck, 15/97, 11 March 2004: 8600). Similarly, force transformation was portrayed as a success. Defense Minister Jung claimed, "Our Bundeswehr has accomplished the transformation process from a pure defense army to an army of unity to an army on operation for peace in an outstanding manner" (16/227, 18 June 2009: 25169). At this point, force transformation had become something that no longer needed justification.

Second, the hegemonization of networked security (and out-of-area operations with it) was increasingly institutionalized in the form of policy documents. Although policy documents are not legally binding, they represent a form of self-commitment that, in case of noncompliance, invites criticism by academics, interest groups, and the media. As I have argued in the previous chapters, the emergence of the project of networked security was a gradual process, and elements of what would become the project of comprehensive (and, later, networked) security were already (partially and temporarily) fixed in policy documents before the project fully formed. Thus, articulations of the new threats can already be found in the 1994 white paper on defense (still published by the Kohl government), as can be a cautious argument for a more active German foreign policy (Federal Ministry of Defence 1994). During chancellor Schröder's term (1998–2005), elements of what was then still called "comprehensive security" were declared to be official government policy in a number of policy documents, most notably the 2000 comprehensive concept (Federal Government 2000) and the 2004 action plan on civilian conflict prevention (Federal Government 2004), the 2003 Defence Policy Guidelines (Federal Ministry of Defence 2003), and the 2004 Conception of the Bundeswehr (BMVg 2004).

Networked security was also inscribed into policy documents issued by the grand coalition government under the leadership of Chancellor Merkel. Most notably, the 2006 white paper on defense lists the action plan for civilian conflict prevention as an "example" for networked security (Federal Ministry of Defence 2006: 23), officially adopting the SPD/Green coalition's efforts to implement comprehensive security (see also Kolbow, 16/60, 26 October 2006: 5788). Through such intertextual constructions of equivalence between different policy papers, the discursive formation is further stabilized. Equally, the 2011 Defence Policy Guidelines and following documents stress a broad understanding of security threats and call for tackling them with a whole-of-government approach, within which a restructured Bundeswehr plays an important role (Federal Ministry of Defence 2011: 2–3, 5). Similar statements can be found in the paper on the restructuring of the Bundeswehr (BMVg 2012a), the report on its implementation (BMVg 2013a), the 2013 Conception of the Bundeswehr (BMVg 2013b), and the 2017 Guidelines on Preventing Crises, Resolving Conflicts, Building Peace (Federal Government 2017), as well as in information brochures like The Bundeswehr on Operations (Federal Ministry of Defence 2009).

Similar developments of institutionalization can be seen on an international level, with the publication of corresponding strategy documents by the EU (2003) and NATO (1999, 2010) as well as the 2003 European Security Strategy,[35] documents regularly cited, in turn, in core German policy documents (e.g., Federal Government 2004: 12, 8; Federal Ministry of Defence 2006: 15, 26–29, 33–35; 2011: 7). Even more recently, the 2016 white paper promotes what it calls a *vernetzter Ansatz* (literally, "networked approach," but translated as "comprehensive approach": Federal Ministry of Defence 2016: 58), which similarly rests on a whole-of-government approach and coordination with nonstate actors (BMVg 2016: 58–59). Equally, more general documents like coalition treaties reinscribe networked security into the discourse (e.g., CDU et al. 2013, 2018).

Third, court rulings and legal codification are examples for sedimentation. An important example in this context are the BVerfG's 1993 and 1994 decisions that (performatively) declared out-of-area operations constitutional (i.e., articulated them as equivalent with the demand for constitutionality) if conducted within systems of collective security and if endowed with a constitutive Bundestag mandate (BVerfGE 89, 38; 90, 286). Equally, force transformation has been accompanied by a number of changes in legislation. As codified binding legal norms, acts of parliament represent a significant step toward the sedimentation and normalization of discursive practices. A number of existing laws have been adapted to accommodate out-of-area operations specifically, including the 1995 Military Law Amendment Act (Wehrrechtsänderungsgesetz, BGBl. I 1995/65), the 2001

---

35. This correspondence is not surprising, given that core documents within NATO and the EU are a product of extensive international coordination.

Bundeswehr Realignment Act (Bundeswehrneuausrichtungsgesetz, BGBl. I 2001/75), the 2005 Parliamentary Participation Act (Parlamentsbeteiligungsgesetz, BGBl. I 2005/17) that regulates the role of the Bundestag in deciding military deployments, the 2011 amendment of the Military Law Amendment Act (BGBl. I 2011/19), the 2012 Bundeswehr Reform Accompanying Act (Bundeswehrreform-Begleitgesetz, BGBl. I 2012/35), the 2004 Operations Care and Provision Act (Einsatzversorgungsgesetz, BGBl. I 2004/72), and the 2012 Act on the Continued Employment of Personnel Injured on Operations (Einsatz-Weiterverwendungsgesetz, BGBl. I 2012/46) (see Raap 1996, 2002; Rau 2006; Schafranek 2009).

Fourth, the project was accompanied by the foundation of new institutions and reform of old ones. Within the Bundestag, networked security is reflected in institutional adaptation, as in the formation, in the 17th legislative period, of the Subcommittee on Civilian Crisis Prevention and Networked Security (from the 18th legislative period onward, the Subcommittee on Civilian Crisis Prevention, Conflict Management, and Networked Action).[36] The project equally entailed changes on the operational level, most notably the creation of Provincial Reconstruction Teams in Afghanistan, which combined military personnel with civilian representatives of the BMZ, the Foreign Office, and the Federal Ministry of the Interior (see Chiari 2013; Ehrhart 2011; Neureuther 2012). Also, force transformation, ongoing since the mid-1990s, is an example for sedimentation. Since the publication of the 1994 white paper, successive governments have undertaken various attempts at restructuring the Bundeswehr to adapt it to out-of-area operations (Deitelhoff and Geis 2008; von Bredow 2015), with the help of a number of expert commissions (see Jacobsen and Rautenberg 1991; von Weizsäcker 2000; Weise et al. 2010).

The most far-reaching reform process yet is the currently ongoing "realignment" (Neuausrichtung) of the Bundeswehr (see BMVg 2012a), initiated by then Defense Minister zu Guttenberg in 2010 and since continued by his successors, de Maizière, Ursula von der Leyen, and, most recently, Annegret Kramp-Karrenbauer (von Bredow 2015). In 2011, after decades of political struggle, conscription was suspended (Harnisch and Weiß 2014: 225–26). Examples for institutional innovation include the establishment of an Operations Command (Meusch 2008), a Center for the Transformation of the Bundeswehr (von Bredow 2015), the Special Operations Division and the

---

36. https://www.bundestag.de/ausschuesse/a03/ua_zks

Airmobile Operations Division (Wiesner 2011: 91),[37] and the Kommando Spezialkräfte, the Bundeswehr's special forces (Deitelhoff and Geis 2008). In 2012, the inspector general, a four-star general and the highest-ranking soldier of the Bundeswehr, previously limited to an advisory capacity (Young 1996: 379), has been installed as the commanding officer of all German soldiers relating to service in military line units (BMVg 2012b).

Another striking example of institutionalization is the 2008 endowment of the Cross of Honor for Valor (Ehrenkreuz für Tapferkeit) and the 2010 establishment of the Combat Action Medal (Einsatzmedaille Gefecht) (BMVg 2012c; critical, Bergmann 2010). The cross, which was, like the combat medal, established to recognize acts of valor in the context of out-of-area operations particularly, can be awarded for "extraordinarily brave deeds" (BMVg 2012c: 4, 12, 18–19).[38] The establishment of a combat medal is particularly striking since German decision makers continue to insist that Germany does not involve itself in wars.[39]

Sedimentation also manifests itself in quite material ways, such as through changed procurement priorities in regard to weapons systems and equipment. For example, force transformation was accompanied by a much stronger focus, within the army (Deutsches Heer), on infantry (although we see a swing back toward the procurement of tanks in the wake of the Ukraine crisis: see BMVg 2016), including the procurement of the armored infantry fighting vehicle Boxer, the project "Infantryman of the Future," unmanned aerial vehicles like the (infamous) Euro Hawk, and the transport helicopter NH90 (Becker 2005; International Institute of Strategic Studies 2007: 101; Schimpf 2012; Schneider and Ritter 2012: 35). Procurement plans for the German Navy, however, reflect an only partly and reluctantly interventionist outlook. While the multipurpose frigate class F125, which integrates a special forces contingent and provides tactical fire support for ground intervention forces, is geared towards out-of-area-operations, the German Navy remains reluctant to procure vessels specialized in amphibious operations (Becker 2005; Stöhs 2018). Nonetheless, overall these quite material processes of institutionalization clearly indicate the extent to which the new threats and a largely interventionist approach to security policy have become accepted.

---

37. Parts of these two divisions have been restructured again to build a new Quick Forces Division (Schneider and Ritter 2012: 35).

38. Theodor Heuss, the first federal president, fundamentally opposed military medals (Becker 2017).

39. I thank an anonymous reviewer for pointing me to this particular irony.

CONCLUSION

This chapter has discussed the transformation, following 9/11, of the project of comprehensive security into what I have called the "project of networked security," which became institutionalized as the new German security order. A number of points are worth revisiting here. First, in contrast to (primarily realist) US security discourse, which was disrupted by 9/11, the German discourse of comprehensive security provided a ready framework for making sense of the terrorist attacks. Moreover, comprehensive security provided a blueprint for post-9/11 security policy. As a result, the discourse of comprehensive, or networked, security emerged as the new security order.

Second, a number of factors contributed to the project's overall appeal. Aside from an exceptionally broad range of equivalent demands (further expanded through the incorporation of security after 9/11), a clear antagonistic frontier, and the provision of an empty signifier as a symbol for the common good, a project's credibility is particularly crucial (see also Nabers 2015; Nabers and Stengel 2019b). In this context, the chapter has highlighted the importance of gendered and racialized discourses. The representative function of a signifier does not necessarily have to be a single signifier. Here, the notion of a blurred empty signifier has proven to be highly useful.

Third, the project of networked security illustrates how identity formation and securitization are linked. While securitization implies the construction of a (threatened) Self, identity formation only requires the exclusion of a radical Other blocking the Self's identity, not necessarily its articulation as a physical security threat. If the identity-security nexus is to be understood, disentangling identity formation and securitization seems crucial.

Fourth, along with the implementation of a new security order went a rearticulation of out-of-area operations form primarily a means to live up to German responsibility to an important instrument to guarantee—if within a broader networked strategy—German security under (supposedly) radically changed international circumstances. At the same time, the incorporation of military means illustrates that securitization can be much more complicated and contradictory than securitization theory would lead one to expect (also Zimmermann 2017), as extraordinary means can sometimes, depending on the relevant set of sedimented practices, only be incorporated into an equivalential chain in a rather unstable fashion. More generally, an analysis of the German security discourse reveals that hegemony is never complete but always contested.

Conclusion

The starting point of this book was the empirical puzzle of the German out-of-area consensus, a development that previous studies have had difficulty explaining satisfactorily. Over the past 30 years, Germany's security policy has undergone a significant transformation away from military restraint and toward what Steele calls "actionism" (2019: 2), with German participation in multinational military operations increasing both quantitatively and qualitatively in terms of the level of intensity. This tendency seems unbroken, with Defense Minister Kramp-Karrenbauer demanding in November 2019 that Germany assume a more active military role in international affairs, bolstered by a bump in defense spending to 2 percent of the German GDP and institutionally supported by a to-be-created national security council (Sprenger 2019).

As I discussed at the outset of the book, despite the widespread claim, among decision makers and researchers alike, that this development presents a necessary adaptation to objective problem pressures (a changed security environment), a closer look reveals it to be far from self-evident. This development is puzzling given a still widespread antimilitarist culture that one would expect to provide a formidable obstacle to German participation in military operations, particularly combat missions such as ISAF. It is even less explicable if one considers the track record of military operations, which mostly do not inspire confidence in terms of their effectiveness beyond UN peace operations, in particular traditional peacekeeping.[1] Against this background, it is difficult to understand the continued trend toward increasing military involvement abroad. Even if one were to accept the somewhat sim-

---

1. See, e.g., Bellamy 2015; Bliesemann de Guevara and Kühn 2010; Downes and Monten 2013; Gilligan and Sergenti 2008; Grimm 2008; Gromes and Dembinsky 2013; Mac Ginty 2012; Ramsbotham et al. 2011; Regan and Meachum 2014.

plistic claim that problems demand solutions (which completely ignores social construction as a factor), it makes little to no sense to argue that they demand largely ineffective ones.

The book uses Essex School discourse theory to provide an explanation for this development. From that perspective, changes in policy (increased participation in military operations) can only be understood if looked at in their wider discursive contexts. In the German case more specifically, the out-of-area consensus that emerged after unification can only be understood if analyzed in the context of much broader discursive change—namely, the transformation of the German security discourse. Rather than seeing Germany's increased participation in military operations as a necessary adaptation to an objectively changed reality, the book zooms in on the processes by which a particular and partial representation of said reality became widely accepted, thus making certain policy responses appear necessary and without alternative. More specifically, this book traces the replacement, over the past 30 years, of the Cold War security order by the discourse of networked security. The Cold War security order focused on the communist "East" as the central security threat to Germany and its allies. The Bundeswehr's main purpose within the German grand strategy during the Cold War was to maintain a conventional deterrent and thus prevent war from ever occurring. In contrast, the discourse of networked security identified new, globalized threats—for example, armed conflict, terrorism, mass migration, and environmental problems—as the most important security challenge after the end of the Cold War. To counter the new threats, Germany had to employ a networked security policy, combining different actors' military and civilian means in a unified strategy and tackling new threats early on and at their place of origin. Military operations were articulated as an indispensable component within such a networked security policy. The central assumption here is that there is a binary choice: either Germany commits to a more interventionist security policy (broadly understood), or it can passively wait for the new threats to reach Germany's shores.

The book not only traces this shift but provides an explanation for how this particular understanding managed to establish itself as the only legitimate, rational way for German parliamentarians to understand the post–Cold War security environment. To explain how some discourses manage to be effective, garnering widespread acceptance, while others fail, a discourse theoretical approach highlights the importance of the construction of a broad equivalential chain of previously disparate social demands, the clear identification of a radical Other standing in the way of realizing these demands, the provision of a unifying symbol (an empty signifier), and the project's overall credibility against deeply sedimented discursive practices. In many respects, the project of networked security is almost an ideal-typical example of a successful hegemonic project. It enlisted an exceptionally broad range of disparate and (in the case of civilian conflict prevention, e.g.) even contradictory demands, thus broadening the project's appeal. It also clearly identified a culprit, in the new threats, for the continued lack of peace and security after the end of the Cold War, and it formulated, with comprehensive/networked security, an equally unambiguous way forward that not only promised to overcome the most pressing problems but held the promise of creating a perfectly peaceful, safe, and just world. Although part of the project was to enable military operations abroad, including combat missions, it managed to draw on and incorporate central traditions of German foreign and security policy after 1945, including antimilitarism. In addition, it managed to sideline and even swallow up competing projects, most notably civilian conflict prevention, which was originally advocated as an alternative to but became rearticulated as compatible with and even depending on military crisis management. Drawing on feminist and postcolonial scholarship, the book also shows how the project gained additional credibility by making use of established discourses on gender and modernity/coloniality.

A brief discussion of the limitations of this study are in order. First, because it focuses on discursive production, it does not offer any explanation, much less a causal one, of why Germany increasingly participated in military operations after unification. Discourse, as it is understood here, does not cause policy actions but, rather, sets the broad framework within which certain policy options seem more or less appropriate, rational, feasible, and moral to pursue, with certain courses for action appearing self-evident and without alternative, while others are not even entertained as serious possibilities. That framework does not mean that subjects are determined by discourse. Discourses are always dislocated, which not only leaves the subject with some modicum of wiggle room but often outright forces it to make decisions between options, of which usually none is unproblematic. Military interventions are a prime example for such tough decisions on an undecidable terrain (Stengel 2019a; Zehfuss 2018). As a consequence, a discourse theoretical explanation can outline the broad conditions of possibility of individual decisions, but it will unavoidably fall short of providing an explanation for why a certain decision has been made—precisely because the subject is, in Laclau's words, "entirely guileless" in the moment of decision (2014e: 51).

Second, the book's focus is clearly limited to the discourse of external security and only touches on questions of domestic security insofar as they are linked, in the discourse, to external security. I am not saying that the external/internal or international/domestic dichotomy is not unproblematic (Walker 1993) or that the internal and the external could be neatly separated in practice (Browning 2003; Eriksson and Rhinard 2009). However, the two are commonly treated as more or less separate in the discourse itself.

Third, while the book provides an explanation for the establishment of out-of-area explanations as a social practice, it does not claim to explain how individual operations have been made possible. Some missions will likely be primarily justified on "humanitarian" grounds (e.g., Allied Force), others with respect to German and/or Western security (e.g., Operation Enduring Freedom in Afghanistan). The focus of this book is on the justification of out-of-area operations as a means of German security policy, and individual justifications of missions (or the refusal to participate, as in Libya) will not follow a universal blueprint and are likely to differ significantly.

Fourth, because the book focuses on out-of-area operations as a social practice, a number of aspects that usually feature prominently in studies about (in particular, specific) out-of-area operations—alliance solidarity and capability (Kaim 2007), the rearticulation of German history (Schwab-Trapp 2003; Zehfuss 2007), shifts like the departure from the so-called Kohl doctrine (Zehfuss 2002: 129), or the changing position of the Green Party (Brunstetter and Brunstetter 2011)—do not receive detailed attention here. Instead, the book focuses specifically on the linking of different discursive elements to demonstrate the functional aspects of the construction of a hegemonic project, at the expense of exhaustiveness and, to some extent at least, the finer nuances of the overall discourse.[2]

Fifth, the analysis of gender and coloniality in the legitimation of military operations is limited. While I have tried to weave feminist and postcolonial arguments into it, the book's analysis is not primarily postcolonial feminist and unavoidably falls short of what such an analysis could achieve. As a result, while the book shows that gender and coloniality are, in my opinion at least, indispensable as analytical categories to make sense of German mili-

---

2. In that regard, see, in particular, Schwab-Trapp's studies (2002, 2007) on the out-of-area debate of the 1990s and the "war on terror."

tary operations, the need for a more systematic and in-depth analysis along these lines persists.

Sixth, its focus on parliamentary debates means that the book can only make claims regarding the hegemonization of networked security within the German Bundestag, not beyond it. The book proceeds from the assumption that the Bundestag (and not, e.g., the mass media) is the central discursive arena in which German foreign policy is constructed in discursive struggles. It cannot demonstrate, largely for practical reasons, whether and how arguments voiced in the Bundestag are picked up elsewhere.

Finally, the book is limited in regards to the period under investigation, which means that certain developments remain beyond its scope. Thus, while the book seeks to reconstruct how the historical antagonism between the FRG and its authoritarian predecessors has been articulated from the 1980s onward, it does not provide an analysis of how, for example, the myth of the new Germany itself became hegemonic or how rearmament was discursively legitimized in the 1950s. Similarly, the analysis stops before the 2014 Ukraine crisis, which was accompanied by securitization processes in Western countries, including Germany (Stahl et al. 2016). It falls on future research to examine how Russian hybrid warfare was articulated in the German security discourse, for example, as a further incarnation of the new threats or as the return of "old school" great power politics.

With its analysis of the German security discourse, the book contributes to a number of empirical and theoretical debates. To begin with, in providing a theoretical explanation for the out-of-area consensus, the book adds to our understanding of German foreign and security policy in a number of ways. First, most broadly, the book zooms in on meaning-making as a crucial aspect of foreign policy decision-making, an aspect often neglected not just by rationalist approaches but also, paradoxically, by conventional constructivist studies, both of which tend to explain policy change as the result of an adaptation to a changed international reality. By putting meaning-making center stage (also notably Hellmann 1999; Zehfuss 2002), the book not only zooms in on an important aspect of the politics involved in foreign policy decision-making that is often neglected (or only addressed in a metatheoretically inconsistent fashion) in the study of German foreign policy but, equally, reveals the taken-for-granted starting point of many studies (i.e., the assumption that reality is unproblematic) as ideological in the Laclauian sense.

Second, more specifically with respect to conventional constructivist

research on German foreign policy, the book reveals the insufficient conception of change at the heart of many constructivist studies (Baumann 2006).
This shortcoming is closely linked to constructivists' essentialist conceptions of "ideational" factors, like norms or identities, that presume for those
factors a more or less clearly defined content that, once internalized by decision makers, somehow constrains behavior. Because they have to presume
ideational factors as more or less stable, conventional constructivist
approaches have a hard time explaining the apparent waxing and waning of
antimilitarism's constraining effect on policy choices, which does not sit
easily with linear notions of ideational change, let alone static conceptions
of ideational factors (Nonhoff and Stengel 2014).

As becomes clear in the book's discussion of antimilitarism, a poststructuralist perspective can provide a more adequate account of social change,
because it can avoid any essentialist understandings of, for example, antimilitarism. The book makes clear that the relationship between military reticence
and policy action is, at the very least, not one of constraint under all circumstances. As the book shows, arguments expressing support for military reticence can be mobilized to make the case for more (not just less) military
engagement or, more generally, to justify actionism (Steele 2019). Researchers
thus need to pay close attention to exactly how discursive elements are articulated in specific discourses. The book demonstrates the analytical added
value of a poststructuralist approach that, based on anti-essentialist conceptions of meaning, zooms in on how discursive elements like antimilitarism
change their meaning depending on how they are arranged vis-à-vis other
elements. As such, a poststructuralist approach is more suited for the analysis
of seemingly paradoxical politics, in which decision makers proclaim one
thing but do the opposite. This aspect should be of broader interest, in particular (but not only), to students of Japanese foreign policy. Since the end of
the Cold War, Japanese security policy has undergone a process of "normalization" (Hughes 2009; Stengel 2007) that is equally difficult to fully grasp
from a conventional constructivist perspective. As in the German case, scholars concerned with Japanese foreign policy have begun to problematize
essentialist understandings of war and peace (Hagström and Hanssen 2016)
and to analyze, for example, the role of realist arguments in undermining
pacifism's credibility as a viable alternative (Gustafsson et al. 2019).

Third, the book zeroes in on a more general methodological problem
(noted in the introduction) afflicting conventional constructivist studies
concerned with German foreign policy. Although constructivism is built on

the assumption of the "social construction of reality" (Berger and Luckmann 1967), many constructivist studies of German foreign policy treat "reality" as unproblematic, claiming that the weakening of antimilitarist norms is due to, as Baumann and Hellmann (2001: 63) aptly sum up, a "reluctant adaptation to a changing international environment"—as if that environment was directly accessible and not subject to processes of social construction, including academics' constructions. Even if one accepts Risse's fair statement that nothing is "empirically eaten as hot, as it has been theoretically cooked" (Risse 2003), treating reality as a given amounts to an ad hoc abandonment of one's own metatheoretical commitments. Emphasizing ideational factors alone does not make for a constructivist study, at least not a coherent one, and taking the meaning of facts as unproblematic means going back even on much of the early work of Wendt (1995)—a man not known for any post-structuralist fits—where he emphasized the importance of meaning for the relevance of material factors. I argue that discourse theory provides a more (meta)theoretically consistent approach to the study of processes of social (or discursive) construction and avoids falling back on a naive realism and/ or empiricism that is untenable from a constructivist position. In highlighting the importance of reflexivity and the political nature of scholarship, this book points to a potential avenue for how constructivist research on German foreign policy could avoid metatheoretical inconsistencies (see also Weller 2005).

Fourth, the book demonstrates the added value of postcolonial and feminist approaches to the study of German foreign and security policy. Aside from a few exceptions (e.g., Engelkamp and Offermann 2012; Schoenes 2011; Shim and Stengel 2017; Ziai 2010), postcolonial and feminist perspectives remain severely underrepresented in a field that sometimes gives the impression of being theoretically stuck at a mid-1990s level (Nonhoff and Stengel 2014: 42n5). In the context of security and defense policy more specifically, this neglect is particularly problematic given the large number of studies that demonstrate the importance of gender and coloniality for the legitimation of armed forces and military interventions (e.g., Eichler 2014; Masters 2009; Muppidi 2012; Peterson 2010; Shepherd 2006). As the book shows, the hegemonization of networked security is hard to comprehend without taking into account the many ways in which the project draws on gendered behavioral expectations and conceptions of security policy as well as deeply engrained civilizationist representations of the Self and the non-Western Other. At the same time, because the book itself is limited in its engagement with postcolonial feminist arguments, one of the most important lacunae in the study of German foreign policy remains a more systematic and in-depth analysis of the role of gender and coloniality. In that regard, the literature remains severely limited, and a dire need remains for further systematic and in-depth studies of these issues from a decidedly postcolonial feminist perspective.

In addition to German foreign policy, the book makes a number of theoretical arguments in regard to issues in security studies, such as the formulation and change of grand strategy and foreign policy more generally, processes of securitization, and debates around the so-called identity-security nexus. First, an Essex School perspective provides a theoretical framework with which to explain grand strategic change (e.g., Miller 2010; Porter 2018). Recently, scholars concerned with grand strategy have turned their attention to the importance of legitimation and narratives for the formation and implementation of grand strategy (Goddard and Krebs 2015; Krebs 2015; Mitzen 2015). Understood from a discourse theoretical perspective, grand strategy is the product of security discourses; it emerges as the result of the replacement of one security order by another in hegemonic struggles over how best to secure the nation. As this book argues with respect to Germany's networked security policy, the establishment of grand strategies rests, like any other hegemonic project, on the creation of a broad coalition through the construction of a chain of equivalent demands, the identification of a clear obstacle (which, in the case of security discourses, is also constructed as a physical security threat), the provision of a universal remedy to tackle the most important security challenges, and the active integration of sedimented practices such as common national myths.

Second, the book speaks to foreign policy change more generally. Exploring discursive change as a vehicle to add to our understanding of foreign policy change, it should be of interest not only to critical constructivist and poststructuralist research in IR and International Political Sociology but, equally, to foreign policy researchers more generally.[3] As numerous poststructuralist studies have argued, discursive transformation and policy change are intimately bound up with each other (Holland 2011), and understanding how the former works adds to our understanding of the latter. This book makes a contribution to that understanding by analyzing a new case

---

3. Foreign policy change has been explored from a number of different theoretical perspectives. See, inter alia, Barnett 1999; Blavoukos and Bourantonis 2014; Carlsnaes 1993; Gustavsson 1999; Hermann 1990; Subotić 2016; Welch 2005.

previously largely neglected (for notable exceptions, see Zehfuss 2002, 2007). The majority of poststructuralist studies in IR have focused on the United States and the discourse of the "war on terror" specifically (Croft 2006; Evans 2013; Holland 2012; Krebs 2015; Lundborg 2012; Nabers 2015; Solomon 2014). Making the German case study particularly useful in this regard is that it shows how discourses such as that on the "war on terror" have to be rearticulated (or translated, if you will: see Stritzel 2014) to be credible for different audiences in light of contingent sedimented practices.

Third, the formulation and transformation of grand strategy is intimately bound up with processes of threat construction or securitization, as grand strategy unavoidably involves the identification of the most important security threats to be countered. Thus, most basically, a discourse theoretical perspective conceptualizes security threats as the contingent and temporary result of discursive struggles. A discourse theoretical perspective provides a theoretical "model" that can shed light on how some securitizing moves manage to transform themselves into successful securitizations and how securitizations are linked to (demands for) extraordinary means in discourse (see Rothe 2015; Stengel 2019b). At the same time, a discourse theoretical perspective highlights the sometimes precarious, unstable, and even contradictory nature of securitization (Zimmermann 2017). In the German case, in particular, the connection (relation of equivalence) between military operations, on one hand, and peace and security as policy goals as well as civilian measures, on the other, proves to be highly difficult to credibly sustain in light of the sedimented discursive practices that structure German security discourse. Most notably, antimilitarism and widespread assumptions about the adequateness of military operations for addressing policy problems such as climate change or mass migration (i.e., causal beliefs) provide a significant obstacle for such articulations and make constant articulatory work necessary to keep the discourse in place. Linked to securitization and the legitimation of extraordinary means is the larger question of the social (i.e., discursive) construction of (il)legitimacy (Hurrelmann et al. 2009; Schmidtke and Nullmeier 2011; Zürn, et al. 2012), specifically in regard to the use of military violence and, on the flip side, pacifism.[4]

Fourth, an Essex School perspective helps us navigate the pitfalls of the

---

4. On the legitimation of the use of force, see, e.g., Bjola 2005; Nuñez-Mietz 2018; O'Driscoll 2018; Zehfuss 2018. On pacifism in IR, see Hutchings 2017; R. Jackson 2019; the contributions to the first issue of *Global Society* 34 (2020).

identity-security nexus (Bilgic 2014; Cho 2012; Mälksoo 2015; Rumelili 2015), that is, the question of exactly how processes of identity formation and threat construction are linked. That debate, in turn, is closely linked to the literature on ontological security (Kinnvall et al. 2018; Kinnvall and Mitzen 2017; Mitzen 2006; Steele 2008; Subotić 2016). In particular, early poststructuralist studies in IR have highlighted the close links between processes of identity formation and the construction of a threatening outside. Although scholars have also explored different types of Self/Other relationships (e.g., Doty 1996), poststructuralist studies have faced the criticism of conflating the physical and ontological constructions of security threats (Rumelili 2015). In my opinion, discourse theory offers a clearer conceptual framework here, distinguishing social antagonism as an ontological necessity for any identity construction from the contingent articulation, at the ontic level, of physical security threats. Put simply, identity formation can but need not involve the construction of an Other as a physical threat (securitization). The same understanding applies, I argue, to othering (the construction of an Other as somehow inferior) or enmity. Identity formation does, however, require the antagonistic construction of an Other as an obstacle to the realization of equivalent demands and, symbolically, a fully constituted identity. Antagonism, then, is much closer to an ontological than to a physical security threat and much more fundamental than the construction of such threats. Thus, an Essex School perspective offers, in principle, a useful conceptual tool kit for disentangling identity formation from physical threat construction and for more systematically analyzing the identity-security nexus.

Fifth, the exploration of the changing meaning of antimilitarism, in particular, speaks to constructivist debates about norms,[5] especially more recent critical contributions that highlight the ambiguity and contestedness of norms and problematize the politics involved in norm research.[6] Discourse theory provides not only an ontological framework that can capture why norms are notoriously ambiguous and contested (Niemann and Schillinger 2017; Wiener 2014) but also a useful way to think through how certain norms emerge (Rosert 2019), are contested, change, weaken, erode (Rosert and Schirmbeck 2007; Zimmermann and Deitelhoff 2019) and vanish (Panke

---

5. Norm dynamics have been a prominent topic in constructivist research since the 1990s (Finnemore and Sikkink 1998, 2001).

6. See, e.g., Deitelhoff and Zimmermann 2013; Engelkamp and Glaab 2015; Engelkamp et al. 2012, 2013; Epstein 2017; Hofius et al. 2014; Ulbert 2012.

and Petersohn 2012). This book has only briefly touched on these issues in its discussion of antimilitarism, but it lends support for an argument in favor of a reconceptualization of norms, including international ones, as the contingent, temporary, and context-dependent products of ongoing discursive struggles. To my knowledge, the only contribution that explores norm dynamics from a discourse theoretical perspective is Renner's (2013) study of reconciliation. Examining to what extent discourse theory can offer insights into different facets of norm dynamics seems like a fruitful area for further research.

Finally, the present book contributes to our understanding of processes of hegemonization, which is relevant for both IR studies inspired by the Essex School and discourse theory.[7] So far, the majority of empirical studies have highlighted the importance of single signifiers or demands that assume the representation of the totality of equivalent demands (e.g., Nabers 2009; Nonhoff 2006; Wullweber 2014). In contrast to the ideal-typical model of hegemonization, the project of networked security was constructed (like the preceding project of comprehensive security) not around a single empty signifier but around what I have called a "blurred empty signifier," that is, a cluster of signifiers referring to broadly the same demand. This blurriness, I have argued, was crucial in enabling the acceptance of the hegemonic project across party lines, because it allowed discourse participants to accept the project while simultaneously claiming it to be new. In addition, the hegemonization of networked security provides insight into how hegemonization need not necessarily be as straightforward as discourse theory suggests. The hegemonization of networked security departs from the ideal-typical model insofar as its original formulation was much narrower in scope. When the project of comprehensive security emerged, it was focused on the German policy of conflict prevention; only after 9/11 did it expand to include the entire security discourse. In a way, the German discourse on conflict prevention functioned as an "incubator" for what would later emerge as the new security order, including, but not limited to, the German policy of conflict prevention. Moreover, empirical analysis suggests that the hierarchical ordering of different signifiers is more complex than expected. Empirical

---

7. A small but growing literature in IR draws on the Essex School to theorize change in international politics. See Biegoń 2016; Eberle 2019; Herschinger 2011; Nabers 2015; Nymalm 2013; Ostermann 2018; Renner 2013; Rothe 2015; Solomon 2014; Wodrig 2017; Wojczewski 2018; contributions to the second issue of New Political Science vol. 41 (2019).

studies inspired by discourse theory often seem to suggest two basic levels of signifiers: privileged empty (or master) signifiers and (subordinated) equivalent moments in the chain. The example of (counter)terrorism discussed in chapter 5 makes clear that certain signifiers can function as central anchor points that are indispensable for the functioning of a discourse, without assuming the role of an empty signifier in the narrow sense.

In highlighting the importance of gender and coloniality for credibility, this book makes the case for discourse theorists to more explicitly engage with other critical perspectives, such as feminist and postcolonial scholarship (see also Nabers 2015). More often than not and certainly in the case of security policy, the sedimented practices that lend credibility to certain discourses often draw on, for example, more established gendered discourses or those influenced by civilizationism and/or modernity/coloniality. Thus, while discourse theory offers a convincing general "model" of hegemony, it makes sense to combine it with more specific arguments to grasp the discursive patterns on which discourses draw for credibility and why they are convincing in a specific context. That recommendation does not mean that any empirical study necessarily has to focus on gender and coloniality specifically. In the context of "Western" military interventions, one would be hardpressed to provide a compelling explanation without taking gender and coloniality into account, but in different circumstances, scholars might want to draw on other approaches, such as critical military studies (Basham et al. 2015; Stavrianakis and Stern 2018), critical race theory (Crenshaw 2011; Mills 2014), critical whiteness studies (Nayak 2007), research on neoliberalism (Springer et al. 2017) or critical geopolitics (Ó Tuathail 1996), among others. In general, the present study shows that it can be very useful to complement a general discourse theoretical framework with other critical perspectives, depending on the topic under investigation.

Overall, the book makes the case for the added value of a discourse theoretical approach for the study of social change in international relations and foreign policy. Through the notions of equivalence, antagonism, representation, and sedimented practices, discourse theory offers a general model of social change that, in principle, should be applicable to an exceptionally broad range of phenomena of concern to IR scholars. Above all, this study demonstrates that any analysis of foreign policy change, norm dynamics, the erosion of taboos, and the legitimation of controversial policies has to place the discursive construction of reality front and center.

References

Abbott, Andrew. 1990. "Positivism and Interpretation in Sociology: Lessons for Sociologists from the History of Stress Research." *Sociological Forum* 5 (3): 435–58.

Abbott, Andrew. 1997. "On the Concept of Turning Point." *Comparative Social Research* 16: 85–105.

Adams, R. J. Q. 1993. *British Politics and Foreign Policy in the Age of Appeasement, 1935–39*. Basingstoke: Palgrave Macmillan.

Adenauer, Konrad. (1949) 2002. "Regierungserklärung vom 20. September 1949." In *Die großen Regierungserklärungen der deutschen Bundeskanzler von Adenauer bis Schröder*, edited by Klaus Stüwe, 33–47. Opladen: Leske + Budrich.

Adler, Emanuel. 2002. "Constructivism and International Relations." In *Handbook of International Relations*, edited by Walter Carlsnaes, Thomas Risse, and Beth A. Simmons, 95–118. 1st ed. London: Sage.

Agius, Christine, and Dean Keep, eds. 2018. *The Politics of Identity: Place, Space and Discourse*. Manchester: Manchester University Press.

Åhäll, Linda. 2012. "The Writing of Heroines: Motherhood and Female Agency in Political Violence." *Security Dialogue* 43 (4): 287–303.

Åhäll, Linda. 2018. Affect as Methodology: Feminism and the Politics of Emotion. *International Political Sociology* 12 (1): 36–52.

Allison, Graham T., and Morton H. Halperin. 1972. "Bureaucratic Politics: A Paradigm and Some Policy Implications." *World Politics* 24 (1): 40–79.

Anderson, Mary B. 1999. *Do No Harm: How Aid Can Support Peace—or War*. London: Lynne Rienner.

Angermüller, Johannes. 2015. *Why There Is No Poststructuralism in France: The Making of an Intellectual Generation*. London: Bloomsbury.

Anning, Stephen, and M. L. R. Smith. 2012. "The Accidental Pirate: Reassessing the Legitimacy of Counterpiracy Operations." *Parameters* 42 (2): 28–41.

Antoniades, Andreas. 2008. "*Cave! Hic* Everyday Life: Repetition, Hegemony and the Social." *British Journal of Politics and International Relations* 10 (3): 412–28.

Aradau, Claudia, and Jef Huysmans. 2014. "Critical Methods in International Relations: The Politics of Techniques, Devices and Acts." *European Journal of International Relations* 20 (2): 596–619.

Aradau, Claudia, Jef Huysmans, Andrew Neal, and Nadine Voelkner, eds. 2014). *Critical Security Methods: New Frameworks for Analysis*. London: Routledge.

Aristotle. 1943. *Politics*. New York: Modern Library.

Ashley, Richard K. 1981. "Political Realism and Human Interests." *International Studies Quarterly* 25 (2): 204–36.

Ashley, Richard K. 1987. "Foreign Policy as Political Performance." *International Studies Notes* 13 (2): 51–54.

Ashley, Richard K., and R. B. J. Walker. 1990. "Introduction: Speaking the Language of Exile; Dissident Thought in International Studies." *International Studies Quarterly* 34 (3): 259–68.

Åslund, Anders. 1992. "Russia's Road from Communism." *Dædalus* 121 (2): 77–95.

Athanassiou, Cerelia. 2012. "'Gutsy' Decisions and Passive Processes." *International Feminist Journal of Politics* 16 (1): 6–25.

Auchter, Jessica. 2012. "Gendering Terror: Discourses of Terrorism and Writing Woman-as-Agent." *International Feminist Journal of Politics* 14 (1): 121–39.

Autesserre, Séverine. 2017. "International Peacebuilding and Local Success: Assumptions and Effectiveness." *International Studies Review* 19 (1): 114–32.

Avey, Paul C., and Michael C. Desch. 2014. "What Do Policymakers Want from Us? Results of a Survey of Current and Former Senior National Security Decision Makers." *International Studies Quarterly* 58 (2): 227–46.

Baaz, Maria Eriksson, and Maria Stern. 2009. "Why Do Soldiers Rape? Masculinity, Violence, and Sexuality in the Armed Forces in the Congo (DRC)." *International Studies Quarterly* 53 (2): 495–518.

Baaz, Maria Eriksson, and Judith Verweijen. 2018. "Confronting the Colonial: The (Re)production of 'African' Exceptionalism in Critical Security and Military Studies." *Security Dialogue* 49 (1–2): 57–69.

Bach, Jonathan P. G. 1999. *Between Sovereignty and Integration: German Foreign Policy and National Identity after 1989*. New York: St. Martin's.

Bahador, Babak, Jeremy Moses, and William Lafi Youmans. 2018. "Rhetoric and Recollection: Recounting the George W. Bush Administration's Case for War in Iraq." *Presidential Studies Quarterly* 48 (1): 4–26.

Bald, Detlef. 2002. "Die Reform des Militärs in der Ära Adenauer." *Geschichte und Gesellschaft* 28 (2): 204–32.

Balzacq, Thierry. 2015. "The 'Essence' of Securitization: Theory, Ideal Type, and a Sociological Science of Security." *International Relations* 29 (1): 103–13.

Banchoff, Thomas. 1999. "German Identity and European Integration." *European Journal of International Relations* 5 (3): 259–89.

Banta, Benjamin. 2013. "Analysing Discourse as a Causal Mechanism." *European Journal of International Relations* 19 (2): 379–402.

Barkawi, Tarak, and Mark Laffey. 2006. "The Postcolonial Moment in Security Studies." *Review of International Studies* 32 (2): 329–52.

Barkawi, Tarak, and Ketih Stanski, eds. 2012. *Orientalism and War*. New York: Columbia University Press.

Barnett, Michael. 1999. "Culture, Strategy and Foreign Policy Change: Israel's Road to Oslo." *European Journal of International Relations* 5 (1): 5–36.

Baron, Ilan Zvi. 2010. "Dying for the State: The Missing Just War Question?" *Review of International Studies* 36 (1): 215–34.

Barthes, Roland. 1967. "The Death of the Author." *Aspen* 5–6, http://www.ubu.com/aspen/aspen5and6/threeEssays.html#barthes.

Basham, Victoria. 2016. "Gender and Militaries: The Importance of Military Masculinities for the Conduct of State Sanctioned Violence." In *Handbook on Gender and War*, edited by Simona Sharoni, Julia Welland, Linda Steiner, and Jennifer Pedersen, 29–46. Cheltenham: Edward Elgar.

Basham, Victoria M., Aaron Belkin, and Jess Gifkins. 2015. "What Is Critical Military Studies?" *Critical Military Studies* 1 (1): 1–2.

Bates, Stephen R., and Laura Jenkins. 2007. "Teaching and Learning Ontology and Epistemology in Political Science." *Politics* 27 (1): 55–63.

Baumann, Rainer. 2001. "Geman Security Policy within Nato." In *German Foreign Policy since Unification: Theories and Case Studies*, edited by Volker Rittberger, 141–84. Manchester: Manchester University Press.

Baumann, Rainer. 2006. *Der Wandel des deutschen Multilateralismus: Eine diskursanalytische Untersuchung deutscher Aussenpolitik*. Baden-Baden: Nomos.

Baumann, Rainer. 2011. "Multilateralismus: Die Wandlung eines vermeintlichen Kontinuitätselements der deutschen Außenpolitik." In *Deutsche Außenpolitik*, edited by Thomas Jäger, Alexander Höse, and Kai Oppermann, 468–87. Wiesbaden: VS Verlag für Sozialwissenschaften.

Baumann, Rainer, and Gunther Hellmann. 2001. "Germany and the Use of Military Force: 'Total War,' the 'Culture of Restraint' and the Quest for Normality." *German Politics* 10 (1): 61–82.

Baumann, Rainer, Volker Rittberger, and Wolfgang Wagner. 1999. "Macht und Machtpolitik: Neorealistische Außenpolitiktheorie und Prognosen über die deutsche Außenpolitik nach der Vereinigung. " *Zeitschrift für Internationale Beziehungen* 6 (2): 245–86.

Baumann, Rainer, and Frank A. Stengel. 2014. "Foreign Policy Analysis, Globalisation and Non-state Actors: State-Centric After All?" *Journal of International Relations and Development* 17 (4): 489–521.

Becker, Ernst Wolfgang. 2017. "Soldatentum und demokratischer Neubeginn." *Militärgeschichtliche Zeitschrift* 76 (2): 459.

Becker, Timm. 2005. "Fegatte Klasse F 125: Neue Wege für die nächste Fregattengeneration der Marine." *MarineForum* 11: 8–16.

Behnke, Andreas. 2006. "The Politics of *Geopolitik* in Post–Cold War Germany." *Geopolitics* 11 (3): 396–419.

Behnke, Andreas. 2012. "The Theme That Dare Not Speak Its Name: Geopolitik, Geopolitics and German Foreign Policy since Unification." In *The Return of Geopolitics in Europe? Social Mechanisms and Foreign Policy Identity Crises*, edited by Stefano Guzzini, 101–26. Cambridge: Cambridge University Press.

Behnke, Andreas. 2013. *Nato's Security Discourse after the Cold War: Representing the West.* New York: Routledge.

Beiner, Marcus. 2009. *Humanities: Was Geisteswissenschaft macht, und was sie ausmacht.* Berlin: Berlin University Press.

Belkin, Aaron. 2012. *Bring Me Men: Military Masculinity and the Benign Façade of American Empire, 1898–2001.* New York: Columbia University Press.

Bell, Duncan. 2002. "Anarchy, Power and Death: Contemporary Political Realism as Ideology." *Journal of Political Ideologies* 7 (2): 221–39.

Bellamy, Alex J. 2015. "When States Go Bad: The Termination of State Perpetrated Mass Killing." *Journal of Peace Research* 52 (5): 565–76.

Bellamy, Alex J., and Tim Dunne, eds. 2016. *The Oxford Handbook of the Responsibility to Protect.* Oxford: Oxford University Press.

Bellamy, Alex J., and Charles T. Hunt. 2015. "Twenty-First Century UN Peace Operations: Protection, Force and the Changing Security Environment." *International Affairs* 91 (6): 1277–98.

Berenskoetter, Felix. 2010. "Identity in International Relations." In *The International Studies Encyclopedia*, edited by Robert A. Denemark. Oxford: Blackwell, https://www.oxfordreference.com/view/10.1093/acref/9780191842665.001.0001/acref-9780191842665-e-0192.

Berger, Peter L., and Thomas Luckmann. 1967. *The Social Construction of Reality.* London: Penguin.

Berger, Thomas U. 1998. *Cultures of Antimilitarism: National Security in Germany and Japan.* Baltimore, MD: Johns Hopkins University Press.

Berger, Thomas U. 2002. "A Perfectly Normal Abnormality." *Japanese Journal of Political Science* 3 (2): 173–93.

Bergmann, Anna. 2010. "Gewalt und Männlichkeit: Wahrnehmungsmuster des 'Fremden' und des 'Eigenen' in der deutschen Berichterstattung über den Afghanistankrieg." In *Medien—Krieg—Geschlecht*, edited by Martina Thiele, Tanja Thomas, and Fabian Virchow, 153–72. Wiesbaden: VS Verlag für Sozialwissenschaften.

Bernstein, Richard. 1983. *Beyond Objectivism and Relativism: Science, Hermeneutics, and Praxis.* Philadelphia: University of Pennsylvania Press.

Bertrand, Sarah. 2018. "Can the Subaltern Securitize? Postcolonial Perspectives on Securitization Theory and Its Critics." *European Journal of International Security* 3 (3): 281–99.

Bevir, Mark. 2003. "Interpretivism: Family Resemblances and Quarrels." *Qualitative Methods* 1 (2): 18–21.

Bevir, Mark. 2008. "Meta-Methodology: Clearing the Underbrush." In *The Oxford Handbook of Political Methodology*, edited by Janet M. Box-Steffensmeier, Henry E. Brady, and David Collier, 48–70. Oxford: Oxford University Press.

Bevir, Mark, and Oliver Daddow. 2015. "Interpreting Foreign Policy: National, Comparative and Regional Studies." *International Relations* 29 (3): 273–87.

Bevir, Mark, and R. A. W. Rhodes. 2003. *Interpreting British Governance.* London: Routledge.

Bhabha, Homi. 1984. "Of Mimicry and Man: The Ambivalence of Colonial Discourse." *October* 28: 125–33.

Bially Mattern, Janice. 2005. *Ordering International Politics: Identity, Crisis, and Representational Force*. London: Routledge.

Biebricher, Thomas, and Eric Vance Johnson. 2012. "What's Wrong with Neoliberalism?" *New Political Science* 34 (2): 202–11.

Biegoń, Dominika. 2016. *Hegemonies of Legitimation: Discourse Dynamics in the European Commission*. London: Palgrave Macmillan.

Biehl, Heiko, Bastian Giegerich, and Alexandra Jonas. 2013. "Introduction." *Strategic Cultures in Europe*, edited by Heiko Biehl, Bastian Giegerich, and Alexandra Jonas, 7–17. Wiesbaden: Springer VS.

Bigo, Didier, and Emma McCluskey. 2018. "What Is a PARIS Approach to (In)securitization? Political Anthropological Research for International Sociology." In *The Oxford Handbook of International Security*, edited by Alexandra Gheciu and William C. Wohlforth. Oxford: Oxford University Press, https://doi.org/10.1093/oxfordhb/9780198777854.013.9.

Bilgic, Ali. 2014. "Trust in World Politics: Converting 'Identity' into a Source of Security through Trust-Learning." *Australian Journal of International Affairs* 68 (1): 36–51.

bin Laden, Osama. 2002. "Letter to the American People." GlobalSecurity.org, http://www.globalsecurity.org/security/library/report/2002/021120-ubl.htm.

Bjola, Corneliu. 2005. "Legitimating the Use of Force in International Politics: A Communicative Action Perspective." *European Journal of International Relations* 11 (2): 266–303.

Bjola, Corneliu, and Markus Kornprobst. 2007. "Security Communities and the Habitus of Restraint: Germany and the United States on Iraq." *Review of International Studies* 33 (2): 285–305.

Blaikie, Norman. 2004a. "Deduction." In *The Sage Encyclopedia of Social Science Research Methods*, edited by Michael S. Lewis-Beck, Alan Bryman, and Tim Futing Liao, 243. Thousand Oaks, CA: Sage.

Blaikie, Norman. 2004b. "Falsificationism." In *The Sage Encyclopedia of Social Science Research Methods*, edited by Michael S. Lewis-Beck, Alan Bryman, and Tim Futing Liao, 377–78. Thousand Oaks, CA: Sage.

Blaney, David L., and Naeem Inayatullah. 2018. "Liberal International Political Economy as Colonial Science." In *The Sage Handbook of the History, Philosophy and Sociology of International Relations*, edited by Andreas Gofas, Inanna Hamati-Ataya, and Nicholas Onuf, 60–74. London: Sage.

Blavoukos, Spyros, and Dimitris Bourantonis. 2014. "Identifying Parameters of Foreign Policy Change: An Eclectic Approach." *Cooperation and Conflict* 49 (4): 483–500.

Bleiker, Roland. 2001. "The Aesthetic Turn in International Political Theory." *Millennium: Journal of International Studies* 30 (3): 509–33.

Bleiker, Roland. 2009. *Aesthetics and World Politics*. Basingstoke: Palgrave Macmillan.

Bleiker, Roland, ed. 2017. *Visual Global Politics*. London: Routledge.

Bleiker, Roland, and Morgan Brigg. 2010. "Introduction to the RIS Forum on Autoethnography and International Relations." *Review of International Studies* 36 (3): 777–78.

Bliesemann de Guevara, Berit, and Florian P. Kühn. 2010. *Illusion Statebuilding: Warum sich der westliche Staat so schwer exportieren lässt.* Hamburg: Edition Körber-Stiftung.

Bliesemann de Guevara, Berit, and Florian P. Kühn. 2011. "'The International Community Needs to Act': Loose Use and Empty Signalling of a Hackneyed Concept." *International Peacekeeping* 18 (2): 135–51.

BMVg. 2004. *Grundzüge der Konzeption der Bundeswehr*. Berlin: BMVg.

BMVg. 2012a. *Die Neuausrichtung der Bundeswehr: Nationale Interessen wahren—Internationale Verantwortung übernehmen—Sicherheit gemeinsam gestalten.* Berlin: BMVg.

BMVg. 2012b. *Dresdner Erlass*. Berlin: BMVg.

BMVg. 2012c. *Ehrenzeichen und Einsatzmedaillen*. Berlin: BMVg.

BMVg. 2013a. *Bericht zum Stand der Neuausrichtung der Bundeswehr.* Berlin: BMVg.

BMVg. 2013b. *Konzeption der Bundeswehr*. Berlin: BMVg.

BMVg. 2016. *Weißbuch 2016 zur Sicherheitspolitik und zur Zukunft der Bundeswehr.* Berlin: BMVg.

Boekle, Henning, Volker Rittberger, and Wolfgang Wagner. 2001. "Constructivist Foreign Policy Theory." In *German Foreign Policy since Unification: Theories and Case Studies*, edited by Volker Rittberger, 105–37. Manchester: Manchester University Press.

Bohnsack, Ralf. 2014. *Rekonstruktive Sozialforschung: Einführung in qualitative Methoden.* 9th ed. Opladen: Leske + Budrich.

Booth, Ken, and Tim Dunne, eds. 2002. *Worlds in Collision: Terror and the Future of Global Order*. Basingstoke: Palgrave Macmillan.

Bosold, David, and Christian Achrainer. 2011. "Die normativen Grundlagen deutscher Außenpolitik." In *Deutsche Außenpolitik*, edited by Thomas Jäger, Alexander Höse, and Kai Oppermann, 445–67. Wiesbaden: VS Verlag für Sozialwissenschaften.

Bourdieu, Pierre. 1984. *Distinction: A Social Critique of the Judgment of Taste*. Cambridge, MA: Harvard University Press.

Bowen, Glenn A. 2006. "Grounded Theory and Sensitizing Concepts." *International Journal of Qualitative Methods* 5 (3): 12–23.

Boyle, Michael J. 2016. "The Coming Illiberal Order." *Survival* 58 (2): 35–66.

Bracher, Karl Dietrich. 1964. *Die Auflösung der Weimarer Republik: Eine Studie zum Problem des Machtverfalls in der Demokratie.* Stuttgart: Ring-Verlag.

Brady, Henry E., and David Collier. 2004. *Rethinking Social Inquiry: Diverse Tools, Shared Standards.* Lanham, MD: Rowman and Littlefield.

Brenke, Gabriele. 1994. "Die Außenpolitik der Bundesrepublik Deutschland." In *Die Internationale Politik, 1991–1992*, edited by Wolfgang Wagner, Marion Gräfin Dönhoff, Lutz Hoffmann, Karl Kaiser, Werner Link, and Hanns W. Maull, 121–32. Munich: R. Oldenbourg.

Brigg, Morgan, and Roland Bleiker. 2010. "Autoethnographic International Relations: Exploring the Self as a Source of Knowledge." *Review of International Studies* 36 (3): 779–98.

Brighi, Elisabetta, and Christopher Hill. 2008. "Implementation and Behaviour." In *Foreign Policy: Theories, Actors, Cases*, edited by Steve Smith, Amelia Hadfield, and Tim Dunne, 117–35. Oxford: Oxford University Press.

Brockmann, Hilke. 1994. "Das wiederbewaffnete Militär: Eine Analyse der Selbstdarstellung der Bundeswehr zwischen 1977 und 1994." *Soziale Welt* 45 (3): 279–303.

Brockmeier, Sarah. 2013. "Germany and the Intervention in Libya." *Survival* 55 (6): 63–90.

Brockmeier, Sarah, and Philipp Rotmann. 2018. "Germany's Politics and Bureaucracy for Preventing Atrocities." *Genocide Studies and Prevention: An International Journal* 11 (3): 20–31.

Brodocz, André. 2009. *Die Macht der Judikative.* Wiesbaden: VS Verlag für Sozialwissenschaften.

Browning, Christopher S. 2003. "The Internal/External Security Paradox and the Reconstruction of Boundaries in the Baltic: The Case of Kaliningrad." *Alternatives: Global, Local, Political* 28 (5): 545–81.

Browning, Christopher S., and Matt McDonald. 2013. "The Future of Critical Security Studies: Ethics and the Politics of Security." *European Journal of International Relations* 19 (2): 235–55.

Brummer, Klaus. 2011. "Überzeugungen und Handeln in der Außenpolitik: Der Operational Code von Angela Merkel und Deutschlands Afghanistanpolitik." *Zeitschrift für Außen- und Sicherheitspolitik* 4 (S1): 143–69.

Brummer, Klaus. 2012. "Germany's Participation in the Kosovo War: Bringing Agency Back In." *Acta Politica* 47 (3): 272–91.

Brummer, Klaus. 2013. "The Reluctant Peacekeeper: Governmental Politics and Germany's Participation in EUFOR RD Congo." *Foreign Policy Analysis* 9 (1): 1–20.

Brummer, Klaus. 2015. "Auslandseinsätze der Bundeswehr: Zwischen Bündnisverpflichtungen, Parlamentsvorbehalt und öffentlicher Meinung." GWP – *Gesellschaft. Wirtschaft. Politik* 64 (1): 49–60.

Brummer, Klaus, and Kai Oppermann. 2019. "Poliheuristic Theory and Germany's (Non-)participation in Multinational Military Interventions: The Noncompensatory Principle, Coalition Politics and Political Survival." *German Politics*, advance online article, https://doi.org/10.1080/09644008.2019.1568992.

Brunner, Claudia. 2017. "Von Selbstreflexion zu Hegemonieselbstkritik." *Sicherheit und Frieden* 35 (4): 196–201.

Brunstetter, Daniel, and Scott Brunstetter. 2011. "Shades of Green: Engaged Pacifism, the Just War Tradition, and the German Greens." *International Relations* 25 (1): 65–84.

Bucher, Bernd. 2015. "Moving Beyond the Substantialist Foundations of the Agency-Structure Dichotomy: Figurational Thinking in International Relations." *Journal of International Relations and Development* 20 (2): 408–33.

Bull, Hedley. 1966. "International Theory: The Case for a Classical Approach." *World Politics* 18 (3): 361–77.

Bulley, Dan. 2010. "The Politics of Ethical Foreign Policy: A Responsibility to Protect Whom?" *European Journal of International Relations* 16 (3): 441–61.

Bundesregierung. 2016. *Weißbuch 2016 zur Sicherheitspolitik und zur Zukunft Der Bundeswehr*. Berlin: Bundesregierung.

Bundestag. 2010. *Basic Law for the Federal Republic of Germany*. Berlin: Deutscher Bundestag.

Bundestag. 2014. *Rules of Procedure of the German Bundestag and Rules of Procedure of the Mediation Committee*. Berlin: Deutscher Bundestag.

Bunge, Mario. 1996. *Finding Philosophy in Social Science*. New Haven, CT: Yale University Press.

Buras, Piotr, and Kerry Longhurst. 2004. "The Berlin Republic, Iraq, and the Use of Force." *European Security* 13 (3): 215–45.

Buro, Andreas. 1995. "Weichenstellung zu ziviler Konfliktbearbeitung in Europa." In *Frieden als Zivilisierungsprojekt*, edited by Wolfgang R. Vogt, 73–82. Baden-Baden: Nomos.

Butler, Judith. 1990. *Gender Trouble: Feminism and the Subversion of Identity*. New York: Routledge.

Butler, Judith. 1993. *Bodies That Matter: On the Discursive Limits of "Sex."* New York: Routledge.

Buzan, Barry, Ole Wæver, and Jaap de Wilde. 1998. *Security: A New Framework for Analysis*. London: Lynne Rienner.

Camargo, Ricardo. 2013. "Rethinking the Political: A Genealogy of the 'Antagonism' in Carl Schmitt through the Lens of Laclau-Mouffe-Zizek." *CR: The New Centennial Review* 13 (1): 161–88.

Cambridge Academic Content Dictionary. 2014. Cambridge: Cambridge University Press, https://dictionary.cambridge.org/dictionary/english/duty?q=duty_1.

Cambridge Academic Content Dictionary. 2014. Cambridge: Cambridge University Press,, https://dictionary.cambridge.org/dictionary/english/responsibility?q=responsibility_1.

Campbell, David. 1993. *Politics without Principle: Sovereignty, Ethics, and the Narratives of the Gulf War*. Boulder, CO: Lynne Rienner.

Campbell, David. 1998. *Writing Security: United States Foreign Policy and the Politics of Identity*. Rev. ed. Minneapolis: University of Minnesota Press.

Campbell, David. 2001a. "International Engagements: The Politics of North American International Relations Theory." *Political Theory* 29 (3): 432–48.

Campbell, David. 2001b. "Time Is Broken: The Return of the Past in the Response to September 11." *Theory and Event* 5 (4), https://muse.jhu.edu/article/32643.

Campbell, Susanna, David Chandler, and Meera Sabaratnam. 2011. "Introduction: The Politics of Liberal Peace." In *A Liberal Peace? The Problems and Prac-*

tices of Peacebuilding, edited by Susanna Campbell, David Chandler, and Meera Sabaratnam, 2–10. London: Zed Books.

Carlsnaes, Walter. 1993. "On Analysing the Dynamics of Foreign Policy Change: A Critique and Reconceptualization." *Cooperation and Conflict* 28 (1): 5–30.

Carlsnaes, Walter. 2002. "Foreign Policy." In *Handbook of International Relations*, edited by Walter Carlsnaes, Thomas Risse, and Beth A. Simmons, 331–49. London: Sage.

Carver, Terrell. 2006. "Being a Man." *Government and Opposition* 41 (3): 450–68.

Carver, Terrell. 2008. "The Machine in the Man." In *Rethinking the Man Question: Sex, Gender and Violence in International Relations*, edited by Jane L. Parpart and Marysia Zalewski, 70–86. London: Zed Books.

Cassese, Antonio. 1999. "*Ex Iniuria Ius Oritur*: Are We Moving Towards International Legitimation of Forcible Humanitarian Countermeasures in the World Community?" *European Journal of International Law* 10 (1): 23–30.

Castells, Manuel. 2011. *The Rise of the Network Society: The Information Age; Economy, Society, and Culture*. 2nd ed. Oxford: Wiley-Blackwell.

CATO Institute. 2002. *The Declaration of Independence and the Constitution of the United States of America*. Washington, DC: CATO Institute.

CDU, CSU, and SPD. 2005. *Working Together for Germany—with Courage and Compassion: Coalition Agreement between the CDU, CSU and SPD*. Berlin: Press and Information Office of the Federal Government.

CDU, CSU, and SPD. 2013. *Deutschlands Zukunft gestalten*. Berlin: PD.

CDU, CSU, and SPD. 2018. *Ein neuer Aufbruch für Europa: Eine neue Dynamik für Deutschland; Ein neuer Zusammenhalt für unser Land*. Berlin: CDU.

Celso, Anthony. 2018. "Al Qaeda's Post 9-11 Travails." *Terrorism and Political Violence* 30 (3): 553–61.

Chakkarath, Pradeep. 2015. "Welt- und Menschenbilder: Eine sozialwissenschaftliche Annäherung." *Aus Politik und Zeitgeschichte* 65 (41–42): 3–9.

Chakrabarty, Dipesh. 2000. *Provincializing Europe: Postcolonial Thought and Historical Difference*. Princeton, NJ: Princeton University Press.

Chandler, David. 2004. "The Responsibility to Protect? Imposing the 'Liberal Peace.'" *International Peacekeeping* 11 (1): 59–81.

Chandler, David. 2010. "The Uncritical Critique of 'Liberal Peace.'" *Review of International Studies* 36 (S1): 137–55.

Chandler, David. 2013. "'Human-Centred' Development? Rethinking 'Freedom' and 'Agency' in Discourses of International Development." *Millennium: Journal of International Studies* 42 (1): 3–23.

Chandler, David. 2015. "The R2P Is Dead, Long Live the R2P: The Successful Separation of Military Intervention from the Responsibility to Protect." *International Peacekeeping* 22 (1): 1–5.

Chandra, Uday. 2013. "The Case for a Postcolonial Approach to the Study of Politics." *New Political Science* 35 (3): 479–91.

Charmaz, Kathy. 2003. "Grounded Theory: Objectivist and Constructivist Methods." In *Strategies of Qualitative Inquiry*, edited by Norman K. Denzin and Yvonna S. Lincoln, 249–91. London: Sage.

Checkel, Jeffrey T. 2001. "Why Comply? Social Learning and European Identity Change." *International Organization* 55 (3): 553–88.

Chiari, Bernhard. 2013. "Die Bundeswehr als Zauberlehrling der Politik? Der ISAF-Einsatz und das Provincial Reconstruction Team (PRT) Kunduz 2003 bis 2012." *Militärgeschichtliche Zeitschrift* 72 (2): 317–51.

Cho, Young Chul. 2012. "State Identity Formation in Constructivist Security Studies: A Suggestive Essay." *Japanese Journal of Political Science* 13 (3): 299–316.

Chong, Alan. 2002. "The Post-International Challenge to Foreign Policy: Signposting 'Plus Non-state' Politics." *Review of International Studies* 28 (4): 783–95.

Chowdhry, Geeta, and Sheila Nair. 2004a. "Introduction: Power in a Postcolonial World; Race, Gender, and Class in International Relations." In *Power, Postcolonialism, and International Relations: Reading Race, Gender and Class*, edited by Geeta Chowdhry and Sheila Nair, 1-32. London: Routledge.

Chowdhry, Geeta, and Sheila Nair, eds. 2004b. *Power, Postcolonialism, and International Relations: Reading Race, Gender and Class*. London: Routledge.

Christensen, Wendy M., and Myra Marx Ferree. 2008. "Cowboy of the World? Gender Discourse and the Iraq War Debate." *Qualitative Sociology* 31 (3): 287–306.

Cilliers, Paul. 2005. Complexity, Deconstruction and Relativism. *Theory, Culture and Society* 22 (5): 255–67.

Clemens, Clay. 1993. "Opportunity or Obligation? Redefining Germany's Military Role Outside of Nato." *Armed Forces & Society* 19 (2): 231–51.

Cockburn, Cynthia. 2010. "Militarism and War." In *Gender Matters in Global Politics: A Feminist Introduction to International Relations*, edited by Laura J. Shepherd, 105–15. London: Routledge.

Cohen, Bernard C., and Scott A. Harris. 1975. "Foreign Policy." In *Handbook of Political Science*. Vol. 6, *Policies and Policymaking*, edited by Fred I. Greenstein and Nelson W. Polsby, 381–437. Reading, MA: Addison-Wesley.

Cohn, Carol. 1987. "Sex and Death in the Rational World of Defense Intellectuals." *Signs: Journal of Women in Culture and Society* 12 (4): 687–718.

Cohn, Carol. 1990. "Clean Bombs and Clean Language." In *Women, Militarism, and War: Essays in History, Politics, and Social Theory*, edited by Jean Bethke Elshtain and Shiela Tobias, 33–55. Savage, MD: Rowman and Littlefield.

Cohn, Carol. 1993. "Wars, Wimps, and Women: Talking Gender and Thinking War." In *Gendering War Talk*, edited by Miriam Cooke and Angela Woollacott, 227–46. Princeton, NJ: Princeton University Press.

Collier, David, and James Mahoney. 1996. "Insights and Pitfalls: Selection Bias in Qualitative Research." *World Politics* 49 (1): 56–91.

Collins, Randall. 1981. "On the Microfoundations of Macrosociology." *American Journal of Sociology* 86 (5): 984–1014.

Condra, Luke N., and Jacob N. Shapiro. 2012. "Who Takes the Blame? The Strategic Effects of Collateral Damage." *American Journal of Political Science* 56 (1): 167–87.

Congressional Research Service. 2019. *Afghanistan: Background and U.S. Policy in Brief*. Washington, DC: Congressional Research Service.

Connell, R. W. 2005. *Masculinities*. 2nd ed. Berkeley: University of California Press.

Conrad, Sebastian. 2012. *German Colonialism: A Short History*. Cambridge: Cambridge University Press.

Conze, Eckart. 2010. "Modernitätsskepsis und die Utopie der Sicherheit: NATO-Nachrüstung und Friedensbewegung in der Geschichte der Bundesrepublik." In *Zeithistorische Forschungen / Studies in Contemporary History* 7: 220–39.

Cook, John, Naomi Oreskes, Peter T. Doran, William R. L. Anderegg, Bart Verheggen, Ed W. Maibach, J. Stuart Carlton, et al. 2016. "Consensus on Consensus: A Synthesis of Consensus Estimates on Human-Caused Global Warming." *Environmental Research Letters* 11 (4), https://doi.org/10.1088/1748-9326/11/4/048002.

Cousin, Glynis. 2010. "Positioning Positionality: The Reflexive Turn." In *New Approaches to Qualitative Research: Wisdom and Uncertainty*, edited by Maggi Savin-Baden and Claire Howell Major, 9–18. London: Routledge.

Coward, Martin. 2018. "Against Network Thinking: A Critique of Pathological Sovereignty." *European Journal of International Relations* 24 (2): 440–63.

Cox, Robert W. 1981. "Social Forces, States and World Orders: Beyond International Relations Theory." *Millennium: Journal of International Studies* 10 (2): 126–55.

Crane-Seeber, Jesse. 2017. "Feminisms I've Known and Loved." *borderlands* 16 (2), http://www.borderlands.net.au/vol16no22017/crane-seeber_feminisms.pdf.

Crawford, Beverly. 2010. "The Normative Power of a Normal State: Power and Revolutionary Vision in Germany's Post-Wall Foreign Policy." *German Politics and Society* 28 (2): 165–84.

Crawford, Beverly, and Kim B. Olsen. 2017. "The Puzzle of Persistence and Power: Explaining Germany's Normative Foreign Policy." *German Politics* 26 (4): 591–608.

Crawford, Neta C. 2002. *Argument and Change in World Politics: Ethics, Decolonization, and Humanitarian Intervention*. Cambridge: Cambridge University Press.

Crenshaw, Kimberlé Williams. 2011. "Twenty Years of Critical Race Theory: Looking Back to Move Forward." *Connecticut Law Review* 43 (5): 1253–352.

Creswell, John W. 2003. *Research Design: Qualitative, Quantitative, and Mixed Methods Approaches*. 3rd ed. Thousand Oaks, CA: Sage.

Critchley, Simon. 2004. "Is There a Normative Deficit in the Theory of Hegemony?" In *Laclau: A Critical Reader*, edited by Simon Critchley and Oliver Marchart, 113–22. London: Routledge.

Critchley, Simon. 2012. *Infinitely Demanding: Ethics of Commitment, Politics of Resistance*. London: Verso.

Croft, Stuart. 2006. *Culture, Crisis and America's War on Terror*. Cambridge: Cambridge University Press.

Crome, Hans-Henning. 2007. "The 'Organisation Gehlen' as Pre-History of the Bundesnachrichtendienst." *Journal of Intelligence History* 7 (1): 31–39.

Crossley-Frolick, Katy A. 2013. "Domestic Constraints, German Foreign Policy and Post-Conflict Peacebuilding." *German Politics and Society* 31 (3): 43–75.

Crossley-Frolick, Katy A. 2017. "Revisiting and Reimagining the Notion of Responsibility in German Foreign Policy." *International Studies Perspectives* 18 (4): 443–64.

Daase, Christopher, and Julian Junk. 2012. "Strategische Kultur und Sicherheitsstrategien in Deutschland." *Sicherheit und Frieden* 30 (3): 152–57.

Dalby, Simon. 1988. "Geopolitical Discourse: The Soviet Union as Other." *Alternatives: Global, Local, Political* 13 (4): 415–42.

Dalby, Simon. 1994. "Gender and Critical Geopolitics: Reading Security Discourse in the New World Disorder." *Environment and Planning D: Society and Space* 12 (5): 595–612.

Dalgaard-Nielsen, Anja. 2003. "Gulf War: The German Resistance." *Survival* 45 (1): 99–116.

Dalgaard-Nielsen, Anja. 2006. *Germany, Pacifism and Peace Enforcement*. Manchester: Manchester University Press.

Darby, Phillip. 2009. "Rolling Back the Frontiers of Empire: Practising the Postcolonial." *International Peacekeeping* 16 (5): 699–716.

Dauphinee, Elizabeth. 2010. "The Ethics of Autoethnography." *Review of International Studies* 36 (3): 799–818.

Dauphinee, Elizabeth. 2013a. *The Politics of Exile*. London: Routledge.

Dauphinee, Elizabeth. 2013b. "Writing as Hope: Reflections on the Politics of Exile." *Security Dialogue* 44 (4): 347–61.

David, Marian. 2013. "The Correspondence Theory of Truth." In *The Stanford Encyclopedia of Philosophy*, edited by Edward N. Zalta. Spring ed. Stanford: Stanford University, https://plato.stanford.edu/archives/fall2013/entries/truth-correspondence/.

Davies, Bronwyn, Jenny Browne, Susanne Gannon, Eileen Honan, Cath Laws, Babette Mueller-Rockstroh, and Eva B. Petersen. 2004. "The Ambivalent Practices of Reflexivity." *Qualitative Inquiry* 10 (3): 360–89.

Dawisha, Adeed I. 1980. "Iraq: The West's Opportunity." *Foreign Policy* 41: 134–53.

Day, Christopher, and Kendra L. Koivu. 2019. "Finding the Question: A Puzzle-Based Approach to the Logic of Discovery." *Journal of Political Science Education* 15 (3): 377–86.

Dean, Jonathan. 1987. "Gorbachev's Arms Control Moves." *Bulletin of the Atomic Scientists* 43 (5): 34–40.

Decker, Frank. 2013a. "Das Parteiensystem vor und nach der Bundestagswahl 2013." *Zeitschrift für Staats- und Europawissenschaften* 11 (3): 323–42.

Decker, Frank. 2013b. "Parteiendemokratie im Wandel." In *Handbuch der deutschen Parteien*, edited by Frank Decker and Viola Neu, 21–60. Wiesbaden: Springer VS.

Deitelhoff, Nicole, and Anna Geis. 2008. "Sicherheits- und Verteidigungspolitik als Gegenstand der Policy- und Governance-Forschung." In *Die Zukunft der Policy-Forschung*, edited by Frank Janning and Katrin Toens, 279–96. Wiesbaden: VS Verlag für Sozialwissenschaften.

Deitelhoff, Nicole, and Lisbeth Zimmermann. 2013. "Aus dem Herzen der Finsternis: Kritisches Lesen und wirkliches Zuhören der konstruktivistischen Normenforschung; Eine Replik auf Stephan Engelkamp, Katharina Glaab und Judith Renner." *Zeitschrift für Internationale Beziehungen* 20 (1): 61–74.

Delgado, Richard, and Jean Stefancic. 2017. *Critical Race Theory: An Introduction*. New York: NYU Press.

Denzin, Norman K. 2009. "The Elephant in the Living Room, or Extending the Conversation about the Politics of Evidence." *Qualitative Research* 9 (2): 139–60.

Denzin, Norman K., and Yvonna S. Lincoln, eds. 2003a. *Collecting and Interpreting Qualitative Materials*. London: Sage.

Denzin, Norman K., and Yvonna S. Lincoln, eds. 2003b. *Strategies of Qualitative Inquiry*. London: Sage.

Denzin, Norman K., and Yvonna S. Lincoln. 2005a. "Introduction: The Discipline and Practice of Qualitative Research." In *The Sage Handbook of Qualitative Research*, edited by Norman K. Denzin and Yvonna S. Lincoln, 1–32. London: Sage.

Denzin, Norman K., and Yvonna S. Lincoln, eds. 2005b. *The Sage Handbook of Qualitative Research*. London: Sage.

Denzin, Norman K., Yvonna S. Lincoln, and Linda Tuhiwai Smith, eds. 2008. *Handbook of Critical and Indigenous Methodologies*. Thousand Oaks, CA: Sage.

Der Derian, James, and Michael J. Shapiro, eds. 1989. *International/Intertextual Relations: Postmodern Readings of World Politics*. Lexington, MA: Lexington Books.

de Saussure, Ferdinand. 2011. *Course in General Linguistics*. New York: Columbia University Press.

Diaz-Bone, Rainer. 2006. "Zur Methodologisierung der Foucaultschen Diskursanalyse." *Historical Social Research* 31 (2): 243–74.

Diez, Thomas. 2001. "Europe as a Discursive Battleground: Discourse Analysis and European Integration Studies." *Cooperation and Conflict* 36 (1): 5–38.

Diez, Thomas. 2004. "Europe's Others and the Return of Geopolitics." *Cambridge Review of International Affairs* 17 (2): 319–35.

Diez, Thomas. 2014. "Bedeutungen und Grenzen: Anmerkungen zur Diskursforschung in den deutschsprachigen Internationalen Beziehungen." In *Diskursforschung in den Internationalen Beziehungen*, edited by Eva Herschinger and Judith Renner, 381–98. Baden-Baden: Nomos.

Diez, Thomas, and Vicki Squire. 2008. "Traditions of Citizenship and the Securitisation of Migration in Germany and Britain." *Citizenship Studies* 12 (6): 565–81.

Dill, Janina, and Henry Shue. 2012. "Limiting the Killing in War: Military Necessity and the St. Petersburg Assumption." *Ethics and International Affairs* 26 (3): 311–33.

Dillon, Michael. 1996. *Politics of Security: Towards a Political Philosophy of Continental Thought*. London: Routledge.

Dillon, Michael, and Julian Reid. 2009. *The Liberal Way of War: Killing to Make Life Live*. London: Routledge.

Doig, Alan, and Mark Phythian. 2005. "The National Interest and the Politics of Threat Exaggeration: The Blair Government's Case for War against Iraq." *Political Quarterly* 76 (3): 368–76.

Doornbos, Martin. 2002. "State Collapse and Fresh Starts: Some Critical Reflections." *Development and Change* 33 (5): 797–815.

Doty, Roxanne Lynn. 1993. "Foreign Policy as Social Construction: A Post-Positivist Analysis of U.S. Counterinsurgency Policy in the Philippines." *International Studies Quarterly* 37 (3): 297–320.

Doty, Roxanne Lynn. 1996. *Imperial Encounters: The Politics of Representation in North-South Relations*. Minneapolis: University of Minnesota Press.

Doty, Roxanne Lynn. 1997. "Aporia: A Critical Exploration of the Agent-Structure Problematique in International Relations Theory." *European Journal of International Relations* 3 (3): 365–92.

Doty, Roxanne Lynn. 2004. "Maladies of Our Souls: Identity and Voice in the Writing of Academic International Relations." *Cambridge Review of International Affairs* 17 (2): 377–92.

Doty, Roxanne Lynn. 2010. "Autoethnography: Making Human Connections." *Review of International Studies* 36 (4): 1047–50.

Downes, Alexander B., and Jonathan Monten. 2013. "Forced to Be Free? Why Foreign-Imposed Regime Change Rarely Leads to Democratization." *International Security* 37 (4): 90–131.

Duden. 2019. Berlin: Dudenverlag, https://www.duden.de/rechtschreibung/Verantwortung, https://www.duden.de/rechtschreibung/verpflichten.

Duffield, John S. 1998. *World Power Forsaken: Political Culture, International Institutions, and German Security Policy after Unification*. Stanford: Stanford University Press.

Duffield, John S. 1999. "Political Culture and State Behavior: Why Germany Confounds Neorealism." *International Organization* 53 (4): 765–803.

Duffield, Mark. 2001. *Global Governance and the New Wars: The Merging of Development and Security*. London: Zed Books.

Duffield, Mark. 2005. "Getting Savages to Fight Barbarians: Development, Security and the Colonial Present." *Conflict, Security and Development* 5 (2): 141–59.

Duffield, Mark. 2007. *Development, Security and Unending War: Governing the World of Peoples*. Cambridge: Polity.

Dunn, Kevin C. 2003. *Imagining the Congo: The International Relations of Identity.* Basingstoke: Palgrave Macmillan.

Dunn, Kevin C. 2008. "Interrogating White Male Privilege." In *Rethinking the Man Question: Sex, Gender and Violence in International Relations,* edited by Jane L. Parpart and Marysia Zalewski, 47–68. London: Zed Books.

Dyson, Tom. 2011. "'Condemned Forever to Becoming and Never to Being'"? The Weise Commission and German Military Isomorphism. *German Politics* 20 (4): 545–67.

Dyson, Tom. 2019. "The Challenge of Creating an Adaptive Bundeswehr." *German Politics,* advance online article, https://doi.org/10.1080/09644008.2019.1612369.

Eagleton-Pierce, Matthew. 2011. "Advancing a Reflexive International Relations." *Millennium: Journal of International Studies* 39 (3): 805–23.

Easter, Gerald M. 1997. "Preference for Presidentialism: Postcommunist Regime Change in Russia and the NIS." *World Politics* 49 (2): 184–211.

Eberle, Jakub. 2019. *Discourse and Affect in Foreign Policy: Germany and the Iraq War.* New York: Routledge.

Eberle, Jakub, and Vladimír Handl. 2020. "Ontological Security, Civilian Power, and German Foreign Policy toward Russia." *Foreign Policy Analysis* 16 (1): 41–58.

Eckstein, Harry. 1975. "Case Study and Theory in Political Science." In *Handbook of Political Science.* Vol. 7, *Strategies of Inquiry,* edited by Fred I. Greenstein and Nelson W. Polsby, 79–137. Reading, MA: Addison-Wesley.

Edinger, Lewis J. 1960. "Post-Totalitarian Leadership: Elites in the German Federal Republic." *American Political Science Review* 54 (1): 58–82.

Edkins, Jenny. 1999. *Poststructuralism and International Relations: Bringing the Political Back In.* Boulder, CO: Lynne Rienner.

Edkins, Jenny. 2002. "Forget Trauma? Responses to September 11." *International Relations* 16 (2): 243–56.

Edkins, Jenny. 2013. "Novel Writing in International Relations: Openings for a Creative Practice." *Security Dialogue* 44 (4): 281–97.

Edkins, Jenny, and Nick Vaughan-Williams, eds. 2009. *Critical Theorists and International Relations.* London: Routledge.

Egbering, Christian. 2011. "Friedenspolitik zwischen ziviler Konfliktbearbeitung und Militärintervention." *Vorgänge* 50 (1): 118–28.

Ehrhart, Hans-Georg. 2011. "Zivil-militärisches Zusammenwirken und vernetzte Sicherheit als Herausforderung deutscher Sicherheitspolitik: Der Fall Afghanistan." *Zeitschrift für Außen- und Sicherheitspolitik* 4 (1): 65–85.

Eichler, Maya. 2014. "Militarized Masculinities in International Relations." *Brown Journal of World Affairs* 21 (1): 81–93.

Eilders, Christiane, and Albrecht Lüter. 2000. "Germany at War: Competing Framing Strategies in German Public Discourse." *European Journal of Communication* 15 (3): 415–28.

Elshtain, Jean Bethke. 1982. "On Beautiful Souls, Just Warriors and Feminist Consciousness." *Women's Studies International Forum* 5 (3): 341–48.

Elster, Jon. 1988. "The Nature and Scope of Rational-Choice Explanation." In *Science in Reflection: The Israel Colloquium: Studies in History, Philosophy, and Sociology of Science*, vol. 3, edited by Edna Ullmann-Margalit, 51–65. Dordrecht: Kluwer.

Engelkamp, Stephan, and Katharina Glaab. 2015. "Writing Norms: Constructivist Norm Research and the Politics of Ambiguity." *Alternatives: Global, Local, Political* 40 (3–4): 201–18.

Engelkamp, Stephan, Katharina Glaab, and Judith Renner. 2012. "In der Sprechstunde: Wie (kritische) Normenforschung ihre Stimme wiederfinden kann." *Zeitschrift für Internationale Beziehungen* 19 (2): 101–28.

Engelkamp, Stephan, Katharina Glaab, and Judith Renner. 2013. "Ein Schritt vor, zwei Schritte zurück? Eine Replik auf Nicole Deitelhoff und Lisbeth Zimmermann." *Zeitschrift für Internationale Beziehungen* 20 (2): 105–18.

Engelkamp, Stephan, and Philipp Offermann. 2012. "It's a Family Affair: Germany as a Responsible Actor in Popular Culture Discourse." *International Studies Perspectives* 13 (3): 235–53.

Enloe, Cynthia. 2000. *Maneuvers: The International Politics of Militarizing Women's Lives*. Berkeley: University of California Press.

Enloe, Cynthia. 2014. *Bananas, Beaches and Bases: Making Feminist Sense of International Politics*. 2nd revised ed. Berkeley, CA: University of California Press.

Enloe, Cynthia. 2016. *Globalization and Militarism: Feminists Make the Link*. 2nd ed. Lanham, MD: Rowman and Littlefield.

Enskat, Sebastian, and Carlo Masala. 2015. "Einsatzarmee Bundeswehr: Fortsetzung der deutschen Außenpolitik mit anderen Mitteln?" *Zeitschrift für Außen- und Sicherheitspolitik* 8 (1): 365–78.

Entman, Robert M. 2003. "Cascading Activation: Contesting the White House's Frame after 9/11." *Political Communication* 20 (4): 415–32.

Epstein, Charlotte. 2011. "Who Speaks? Discourse, the Subject and the Study of Identity in International Politics." *European Journal of International Relations* 17 (2): 327–50.

Epstein, Charlotte, ed. 2017. *Against International Relations Norms: Postcolonial Perspectives*. London: Routledge.

Eriksson, Johan, and Mark Rhinard. 2009. "The Internal-External Security Nexus: Notes on an Emerging Research Agenda." *Cooperation and Conflict* 44 (3): 243–67.

EU. 2003. *A Secure Europe in a Better World: European Security Strategy*. Brussels: European Union, High Representative for the Common Foreign and Security Policy.

Evans, Brad. 2013. *Liberal Terror*. Cambridge: Polity.

Fairclough, Norman. 2013. "Critical Discourse Analysis and Critical Policy Studies." *Critical Policy Studies* 7 (2): 177–97.

Federal Government. 2000. *Comprehensive Concept of the Federal Government on Civilian Crisis Prevention, Conflict Resolution and Post-Conflict Peace-Building*. Berlin: Federal Government.

Federal Government. 2004. *Action Plan: "Civilian Crisis Prevention, Conflict Resolution and Post-Conflict Peace-Building."* Berlin: Federal Government.

Federal Government. 2017. *Guidelines on Preventing Crises, Resolving Conflicts, Building Peace.* Berlin: Federal Government.

Federal Ministry of Defence. 1994. *White Paper on the Security of the Federal Republic of Germany and the Situation and Future of the Bundeswehr.* Berlin: Press and Information Office of the Federal Government.

Federal Ministry of Defence. 2003. *Defence Policy Guidelines.* Berlin: Federal Ministry of Defence.

Federal Ministry of Defence. 2006. *White Paper 2006 on German Security Policy and the Future of the Bundeswehr.* Berlin: Federal Ministry of Defence.

Federal Ministry of Defence. 2009. *The Bundeswehr on Operations.* 2nd ed. Berlin: Federal Ministry of Defence.

Federal Ministry of Defence. 2011. *Defence Policy Guidelines.* Berlin: Federal Ministry of Defence.

Federal Ministry of Defence. 2016. *White Paper on German Security Policy and the Future of the Bundeswehr.* Berlin: Federal Ministry of Defence.

Feldman, Regina. 2003. "German by Virtue of Others: The Search for Identity in Three Debates." *Cultural Studies* 17 (2): 250–74.

Figueroa Helland, Leonardo, and Stefan Borg. 2014. "The Lure of State Failure: A Critique of State Failure Discourse in World Politics." *Interventions* 16 (6): 877–97.

Finnemore, Martha, and Kathryn Sikkink. 1998. "International Norm Dynamics and Political Change." *International Organization* 52 (4): 887–917.

Finnemore, Martha, and Kathryn Sikkink. 2001. "Taking Stock: The Constructivist Research Program in International Relations and Comparative Politics." *Annual Review of Political Science* 4 (1): 391–416.

Fischer, Joschka. 1999. *Rede Joschka Fischers auf dem Außerordentlichen Parteitag in Bielefeld, 13.5.99,* https://web.archive.org/web/20170924001517/http://staff-www.uni-marburg.de/~naeser/kos-fisc.htm.

Fischer, Martina. 2004. "Der Aktionsplan Krisenprävention der Bundesregierung: Von der Bestandsaufnahme zur Selbstverpflichtung für eine zivile Außenpolitik?" *Die Friedens-Warte* 79 (3–4): 313–22.

Fleming, Crystal M. 2018. *How to Be Less Stupid about Race: On Racism, White Supremacy, and the Racial Divide.* Boston: Beacon.

Flick, Uwe. 2006. *An Introduction to Qualitative Research.* 3rd ed. Thousand Oaks, CA: Sage.

Flockhart, Trine. 2016. "The Problem of Change in Constructivist Theory: Ontological Security Seeking and Agent Motivation." *Review of International Studies* 42 (5):799–820.

Flügel-Martinsen, Oliver. 2010. "Die Normativität von Kritik: Ein Minimalmodell." *Zeitschrift für Politische Theorie* 1 (2): 139–54.

Fortna, Virginia Page. 2004. "Does Peacekeeping Keep Peace? International Intervention and the Duration of Peace after Civil War." *International Studies Quarterly* 48 (2): 269–92.

Fortna, Virginia Page, and Lise Morje Howard. 2008. "Pitfalls and Prospects in the Peacekeeping Literature." *Annual Review of Political Science* 11 (1): 283–301.

Foucault, Michel. 1971. "Orders of Discourse." *Social Science Information* 10 (2): 7–30.

Foucault, Michel. 1980. *Power/Knowledge: Selected Interviews and Other Writings, 1972–1977*. New York: Pantheon.

Foucault, Michel. 1982. "The Subject and Power." *Critical Inquiry* 8 (4): 777–95.

Foucault, Michel. 2002. *The Archaeology of Knowledge*. London: Routledge.

Franke, Ulrich, and Ulrich Roos. 2010. "Rekonstruktionslogische Forschungsansätze." In *Handbuch der Internationalen Politik*, edited by Carlo Masala, Frank Sauer and Andreas Wilhelm, 285–303. Wiesbaden: VS Verlag für Sozialwissenschaften.

Franke, Ulrich, and Ulrich Roos. 2013a. "Einleitung." In *Rekonstruktive Methoden der Weltpolitikforschung: Anwendungsbeispiele und Entwicklungstendenzen*, edited by Ulrich Franke and Ulrich Roos, 7–58. Baden-Baden: Nomos.

Franke, Ulrich, and Ulrich Roos. 2013b. "Rekonstruktive Methoden der Weltpolitikforschung: Anwendungsbeispiele und Entwicklungstendenzen." Baden-Baden: Nomos.

Franke, Ulrich, and Kaspar Schiltz. 2013. "'They Don't Really Care about Us!': On Political Worldviews in Popular Music." *International Studies Perspectives* 14 (1): 39–55.

Franke, Ulrich, and Ralph Weber. 2012. "At the Papini Hotel: On Pragmatism in the Study of International Relations." *European Journal of International Relations* 18 (4): 669–91.

Franklin, M. I., ed. 2005. *Resounding International Relations: On Music, Culture, and Politics*. Basingstoke: Palgrave Macmillan.

Frazer, Elizabeth, and Kimberly Hutchings. 2014. "Revisiting Ruddick: Feminism, Pacifism and Non-violence." *Journal of International Political Theory* 10 (1): 109–24.

Freeden, Michael. 1996. *Ideologies and Political Theory: A Conceptual Approach*. Oxford: Oxford University Press.

Freeden, Michael. 2003. *Ideology: A Very Short Introduction*. Oxford: Oxford University Press.

Freedman, Lawrence. 2004. "War in Iraq: Selling the Threat." *Survival* 46 (2): 7–50.

Friedrichs, Jörg, and Friedrich Kratochwil. 2009. "On Acting and Knowing: How Pragmatism Can Advance International Relations Research and Methodology." *International Organization* 63 (4): 701–31.

Friedrichsmeyer, Sara, Sara Lennox, and Susanne Zantop, eds. 1998. *The Imperialist Imagination: German Colonialism and Its Legacy*. Ann Arbor: University of Michigan Press.

Fukuyama, Francis. 1989. "The End of History?" *National Interest* (Summer 1989): 3–18.

Galtung, Johan, and Tord Hoivik. 1971. "Structural and Direct Violence: A Note on Operationalization." *Journal of Peace Research* 8 (1): 73–76.

Gareis, Sven Bernhard. 2005. *Deutschlands Außen- und Sicherheitspolitik: Eine Einführung.* Opladen: Verlag Barbara Budrich.

Gareis, Sven Bernhard. 2010. "Zwischen Bündnisräson und Parlamentsvorbehalt: Wer entscheidet über Auslandseinsätze der Bundeswehr?" In *Friedensethik und Sicherheitspolitik*, edited by Angelika Dörfler-Dierken and Gerd Portugall, 153–68. Wiesbaden: VS Verlag für Sozialwissenschaften.

Gauck, Joachim. 2014. *Germany's Role in the World: Reflections on Responsibility, Norms and Alliances*, speech at the Munich Security Conference, 31 January 2014, http://www.bundespraesident.de/SharedDocs/Downloads/DE/ Reden/2014/01/140131-Muenchner-Sicherheitskonferenz-Englisch.pdf;jsessi onid=25BD48CF3EFC35F822DDF4853C3 80D20.2_cid388?__blob=publicationFile.

Geddes, Barbara. 2003. *Paradigms and Sandcastles: Theory Building and Research Design in Comparative Politics.* Ann Arbor: University of Michigan Press.

Geertz, Clifford. 1975. "On the Nature of Anthropological Understanding." *American Scientist* 63 (1): 47–53.

Geis, Anna. 2001. "Diagnose: Doppelbefund—Ursache: ungeklärt? Die Kontroversen um den "demokratischen Frieden."" *Politische Vierteljahresschrift* 42 (2): 282–98.

Geis, Anna. 2008. Militär und Friedenspolitik: Dilemmata der deutschen "Zivilmacht." In *Berliner Friedenspolitik? Militärische Transformation—zivile Impulse—europäische Einbindung*, edited by Peter Schlotter, Wilhelm Nolte, and Renate Grasse, 60–82. Baden-Baden: Nomos.

Geis, Anna. 2013a. "Burdens of the Past, Shadows of the Future: The Use of Military Force as a Challenge for the German 'Civilian Power.'" In *The Militant Face of Democracy: Liberal Forces for Good*, edited by Anna Geis, Harald Müller, and Niklas Schörnig, 231–68. Cambridge: Cambridge University Press.

Geis, Anna. 2013b. "The 'Concert of Democracies': Why Some States Are More Equal Than Others." *International Politics* 50 (2): 257–77.

Geis, Anna. 2019. "Warten auf die große sicherheitspolitische Debatte in Deutschland?" In *Das Weißbuch 2016 und die Herausforderungen von Strategiebildung*, edited by Daniel Jacobi and Gunther Hellmann, 199–221. Wiesbaden: Springer VS.

Geis, Anna, and Hanna Pfeifer. 2017. "Deutsche Verantwortung in der 'Mitte der Gesellschaft' aushandeln? Über Politisierung und Entpolitisierung der deutschen Außenpolitik." In *Politik und Verantwortung: Analysen zum Wandel politischer Entscheidungs- und Rechtfertigungspraktiken*, edited by Christopher Daase, Julian Junk, Stefan Kroll, and Valentin Rauer, 218–43. Baden-Baden: Nomos.

Geis, Anna, and Wolfgang Wagner. 2011. "How Far Is It from Königsberg to Kandahar? Democratic Peace and Democratic Violence in International Relations." *Review of International Studies* 37 (4): 1555–77.

Genschel, Philipp, and Bernhard Zangl. 2008. "Metamorphosen des Staates—

vom Herrschaftsmonopolisten zum Herrschaftsmanager." *Leviathan* 36 (3): 430–54.

Genschel, Philipp, and Bernhard Zangl. 2011. "L'état et l'exercice de l'autorité politique: dénationalisation et administration." *Revue française de sociologie* 52 (3): 509–35.

Genschel, Philipp, and Bernhard Zangl. 2013. "State Transformations in OECD Countries." *Annual Review of Political Science* 17 (1): 337–54.

Genschel, Philipp, and Bernhard Zangl. 2017. "The Rise of Non-state Authority and the Reconfiguration of the State." In *Reconfiguring European States in Crisis*, edited by Desmond King and Patrick Le Galès. Oxford: Oxford University Press, https://doi.org/10.1093/acprof:oso/9780198793373.003.0003.

George, Alexander L., and Andrew Bennett. 2005. *Case Studies and Theory Development in the Social Sciences*. Cambridge, MA: MIT Press.

George, Jim. 1989. "International Relations and the Search for Thinking Space: Another View of the Third Debate." *International Studies Quarterly* 33 (3): 269–79.

George, Jim. 1994. *Discourses of Global Politics: A Critical (Re)introduction to International Relations*. Boulder, CO: Lynne Rienner.

Geras, Norman. 1987. "Post-Marxism?" *New Left Review* 163: 40–82.

Gerring, John. 2007. *Case Study Research: Principles and Practices*. Cambridge: Cambridge University Press.

Giegerich, Bastian, and Stéfanie von Hlatky. 2019. "Experiences May Vary: Nato and Cultural Interoperability in Afghanistan." *Armed Forces and Society*, advance online article, https://doi.org/10.1177%2F0095327X19875490.

Gießmann, Hans J. 2004. "Deutsche Außenpolitik sollte gescheite Friedenspolitik sein." *WeltTrends* 12 (43): 41–46.

Giles, Geoffrey J., ed. 1997. *Stunde Null: The End and the Beginning Fifty Years Ago*. Washington, DC: German Historical Institute.

Gilligan, Michael J., and Ernest J. Sergenti. 2008. "Do UN Interventions Cause Peace? Using Matching to Improve Causal Inference." *Quarterly Journal of Political Science* 3 (2): 89–122.

Glaser, Barney G. 2002. "Conceptualization: On Theory and Theorizing Using Grounded Theory." *International Journal of Qualitative Methods* 1 (2): 23–38.

Glaser, Barney G., and Anselm L. Strauss. 2006. *The Discovery of Grounded Theory: Strategies for Qualitative Research*. New Brunswick: Transaction.

Glasze, Georg. 2007a. "The Discursive Constitution of a World-Spanning Region and the Role of Empty Signifiers: The Case of Francophonia." *Geopolitics* 12 (4): 656–79.

Glasze, Georg. 2007b. "Vorschläge zur Operationalisierung der Diskurstheorie von Laclau und Mouffe in einer Triangulation von lexikometrischen und interpretativen Methoden." *Forum Qualitative Sozialforschung* 8 (2), http://www.qualitative-research.net/fqs-texte/2-07/07-2-14-d.htm.

Glatz, Rainer L., Wibke Hansen, Markus Kaim, and Judith Vorrath. 2018. *Die Auslandseinsätze der Bundeswehr im Wandel*. SWP-Studie 7. Berlin: Stiftung Wissenschaft und Politik.

Glennie, Alisdair. 2012. "My Baby Days Are Over, Says Gwyneth, I Can't Face Dealing with Nappies Again." *Daily Mail Online*, 19 August, http://www.daily-mail.co.uk/tvshowbiz/article-2190775/My-baby-days-I-dont-feel-I-dealing-nappies-says-Gwyneth.html.

Glynos, Jason, and David Howarth. 2007. *Logics of Critical Explanation in Social and Political Theory*. London: Routledge.

Goddard, Stacie E., and Ronald R. Krebs. 2015. "Rhetoric, Legitimation, and Grand Strategy." *Security Studies* 24 (1): 5–36.

Goertz, Gary, and James Mahoney. 2012a. "Concepts and Measurement: Ontology and Epistemology." *Social Science Information* 51 (2): 205–16.

Goertz, Gary, and James Mahoney. 2012b. *A Tale of Two Cultures: Qualitative and Quantitative Research in the Social Sciences*. Princeton, NJ: Princeton University Press.

Goetze, Catherine. 2017. *The Distinction of Peace: A Social Analysis of Peacebuilding*. Ann Arbor: University of Michigan Press.

Goldhagen, Daniel Jonah. 1996. *Hitler's Willing Executioners*. New York: Knopf.

Goldstein, Joshua S. 2001. *War and Gender: How Gender Shapes the War System and Vice Versa*. Cambridge: Cambridge University Press.

Goldstein, Judith, and Robert O. Keohane. 1993a. "Ideas and Foreign Policy: An Analytical Framework." In *Ideas and Foreign Policy: Beliefs, Institutions, and Political Change*, edited by Judith Goldstein and Robert O. Keohane, 3–30. Ithaca, NY: Cornell University Press.

Goldstein, Judith, and Robert O. Keohane, eds. 1993b. *Ideas and Foreign Policy: Beliefs, Institutions, and Political Change*. Ithaca, NY: Cornell University Press.

Gordon, Philip H. 1994. "The Normalization of German Foreign Policy." *Orbis* 38 (2): 225–44.

Gowans, Chris. 2012. "Moral Relativism." In *The Stanford Encyclopedia of Philosophy*, edited by Edward N. Zalta. Spring ed. Stanford: Stanford University, http://plato.stanford.edu/archives/spr2012/entries/moral-relativism/.

Gravelle, Timothy B., Jason Reifler, and Thomas J. Scotto. 2017. "The Structure of Foreign Policy Attitudes in Transatlantic Perspective: Comparing the United States, United Kingdom, France and Germany." *European Journal of Political Research* 56 (4): 757–76.

Grimm, Sonja. 2008. "External Democratization after War: Success and Failure." *Democratization* 15 (3): 525–49.

Grint, Keith, and Steve Woolgar. 1992. "Computers, Guns, and Roses: What's Social about Being Shot?" *Science, Technology, and Human Values* 17 (3): 366–80.

Gromes, Thorsten. 2012. "Der Rückfall in den Bürgerkrieg." *Zeitschrift für Friedens- und Konfliktforschung* 1 (2): 275–305.

Gromes, Thorsten, and Matthias Dembinski. 2013. *Bestandsaufnahme der humanitären militärischen Interventionen zwischen 1947 und 2005*. HSFK Report 2. Frankfurt am Main: HSFK.

Grovogui, Siba N. 2010. "Postcolonialism." In *International Relations Theories: Dis-*

cipline and Diversity, edited by Tim Dunne, Milja Kurki, and Steve Smith, 247–65. Oxford: Oxford University Press.

Guillaume, Xavier. 2013. "Criticality." In *Research Methods in Critical Security Studies: An Introduction*, edited by Mark B. Salter and Can E. Mutlu, 29–32. London: Routledge.

Gustafsson, Karl, Linus Hagström, and Ulv Hanssen. 2018. "Japan's Pacifism Is Dead." *Survival* 60 (6): 137–58.

Gustafsson, Karl, Linus Hagström, and Ulv Hanssen. 2019. "Long Live Pacifism! Narrative Power and Japan's Pacifist Model." *Cambridge Review of International Affairs* 32 (4): 502–20.

Gustavsson, Jakob. 1999. "How Should We Study Foreign Policy Change?" *Cooperation and Conflict* 34 (1): 73–95.

Guyer, Paul, and Rolf-Peter Horstmann. 2015. "Idealism." In *The Stanford Encyclopedia of Philosophy*, edited by Edward N. Zalta. Fall ed. Stanford: Stanford University, http://plato.stanford.edu/entries/idealism/.

Guzzini, Stefano. 2000. "A Reconstruction of Constructivism in International Relations." *European Journal of International Relations* 6 (2): 147–82.

Guzzini, Stefano. 2011. "Securitization as a Causal Mechanism." *Security Dialogue* 42 (4–5): 329–41.

Hacke, Christian. 2003. *Die Außenpolitik der Bundesrepublik Deutschland: Von Konrad Adenauer bis Gerhard Schröder*. 1st ed. Munich: Ullstein.

Hagström, Linus, and Ulv Hanssen. 2016. "War Is Peace: The Rearticulation of 'Peace' in Japan's China Discourse." *Review of International Studies* 42 (2): 266–86.

Hagström, Linus, and Erik Isaksson. 2019. "Pacifist Identity, Civics Textbooks, and the Opposition to Japan's Security Legislation." *Journal of Japanese Studies* 45 (1): 31–55.

Hajer, Maarten A. 2005. "Coalitions, Practices, and Meaning in Environmental Politics: From Acid Rain to BSE." In *Discourse Theory in European Politics*, edited by David Howarth and Jacob Torfing, 297–315. Basingstoke: Palgrave Macmillan.

Hall, Stuart. 1992. "The West and the Rest: Discourse and Power. In *Modernity: An Introduction to Modern Societies*, edited by Stuart Hall, David Held, Don Hubert, and Kenneth Thompson, 184–227. Oxford: Blackwell.

Halliday, Fred. 1996. "The Future of International Relations: Fears and Hopes." In *International Theory: Positivism and Beyond*, edited by Ken Booth, Steve Smith, and Marysia Zalewski, 318–27. Cambridge: Cambridge University Press.

Halper, Stefan, and Jonathan Clarke. 2004. *America Alone: The Neo-Conservatives and the Global Order*. Cambridge: Cambridge University Press.

Hamati-Ataya, Inanna. 2011. "The 'Problem of Values' and International Relations Scholarship: From Applied Reflexivity to Reflexivism." *International Studies Review* 13 (2): 259–87.

Hamati-Ataya, Inanna. 2013. "Reflectivity, Reflexivity, Reflexivism: IR's 'Reflexive Turn'—and Beyond." *European Journal of International Relations* 19 (4): 669–94.

Hamati-Ataya, Inanna. 2018. "Crafting the Reflexive Gaze: Knowledge of Knowledge in the Social Worlds of International Relations." In *The Sage Handbook of the History, Philosophy and Sociology of International Relations*, edited by Andreas Gofas, Inanna Hamati-Ataya, and Nicholas Onuf, 13–30. London: Sage.

Hammersley, Martyn. 1992. *What's Wrong with Ethnography? Methodological Explorations*. London: Routledge.

Hanrieder, Tine. 2011. "The False Promise of the Better Argument." *International Theory* 3 (3): 390–415.

Hanrieder, Tine, and Christian Kreuder-Sonnen. 2014. "Who Decides on the Exception? Securitization and Emergency Governance in Global Health." *Security Dialogue* 45 (4): 331–48.

Hansen, Lene. 2006. *Security as Practice: Discourse Analysis and the Bosnian War*. London: Routledge.

Hansen, Lene. 2011. "Theorizing the Image for Security Studies." *European Journal of International Relations* 17 (1): 51–74.

Harnisch, Sebastian. 2001. "Change and Continuity in Post-Unification German Foreign Policy." *German Politics* 10 (1): 35–60.

Harnisch, Sebastian, and Simon Weiß. 2014. "Rapider Politikwechsel in der deutschen Verteidigungspolitik: Eine analytische Kurzgeschichte der Suspendierung der Wehrpflicht." In *Rapide Politikwechsel in der Bundesrepublik: Theoretischer Rahmen und empirische Befunde*, edited by Friedbert W. Rüb, 206–39. Baden-Baden: Nomos.

Haverland, Markus, and Dvora Yanow. 2012. "A Hitchhiker's Guide to the Public Administration Research Universe: Surviving Conversations on Methodologies and Methods." *Public Administration Review* 73 (3): 401–8.

Hawkesworth, Mary. 2010. "From Constitutive Outside to the Politics of Extinction: Critical Race Theory, Feminist Theory, and Political Theory." *Political Research Quarterly* 63 (3): 686–96.

Heck, Axel. 2017. "Analyzing Docudramas in International Relations: Narratives in the Film *A Murderous Decision*." *International Studies Perspectives* 18 (4): 365–90.

Hellmann, Gunther. 1999. "Machtbalance und Vormachtdenken sind überholt: Zum außenpolitischen Diskurs in Deutschland." In *Außenpolitischer Wandel in theoretischer und vergleichender Perspektive*, edited by Monika Medick-Krakau, 97–126. Baden-Baden: Nomos.

Hellmann, Gunther. 2002. "Sag beim Abschied leise servus! Die Zivilmacht Deutschland beginnt, ein neues 'Selbst' zu behaupten." *Politische Vierteljahresschrift* 43 (3): 498–507.

Hellmann, Gunther. 2007. "'. . . um diesen deutschen Weg zu Ende gehen zu können': Die Renaissance machtpolitischer Selbstbehauptung in der zweiten Amtszeit der Regierung Schröder-Fischer." In *Ende des rot-grünen Projektes: Eine Bilanz der Regierung Schröder 2002–2005*, edited by Christoph Egle and Reimut Zolnhöfer, 453–79. Wiesbaden: VS Verlag für Sozialwissenschaften.

Hellmann, Gunther. 2009a. "Beliefs as Rules for Action: Pragmatism as a Theory of Thought and Action." *International Studies Review* 11 (3): 638–41.

Hellmann, Gunther. 2009b. "Fatal Attraction? German Foreign Policy and IR/ Foreign Policy Theory." *Journal of International Relations and Development* 12 (3): 257–92.

Hellmann, Gunther. 2011. "Das neue Selbstbewusstsein deutscher Außenpolitik und die veränderten Standards der Angemessenheit." In *Deutsche Außenpolitik*, edited by Thomas Jäger, Alexander Höse, and Kai Oppermann, 735–57. Wiesbaden: VS Verlag für Sozialwissenschaften.

Hellmann, Gunther. 2016a. "Foreign Policy: Concept, Vocabulary, and Practice." In *The Transformation of Foreign Policy: Drawing and Managing Boundaries from Antiquity to the Present*, edited by Gunther Hellmann, Andreas Fahrmeir, and Miloš Vec, 30–50. Oxford: Oxford University Press.

Hellmann, Gunther. 2016b. "Germany's World: Power and Followership in a Crisis-Ridden Europe." *Global Affairs* 2 (1): 3–20.

Hellmann, Gunther, Andreas Fahrmeir, and Miloš Vec, eds. 2016a. *The Transformation of Foreign Policy: Drawing and Managing Boundaries from Antiquity to the Present*. Oxford: Oxford University Press.

Hellmann, Gunther, Andreas Fahrmeir, and Miloš Vec. 2016b. "The Transformation of Foreign Policy: Legal Framework, Historiography, Theory." In *The Transformation of Foreign Policy: Drawing and Managing Boundaries from Antiquity to the Present*, edited by Gunther Hellmann, Andreas Fahrmeir, and Miloš Vec, 13–29. Oxford: Oxford University Press.

Hellmann, Gunther, and Benjamin Herborth, eds. 2016. *Uses of "the West": Security and the Politics of Order*. Cambridge: Cambridge University Press.

Hellmann, Gunther, Christian Weber, Frank Sauer, and Sonja Schirmbeck. 2007. "'Selbstbewusst' und 'stolz': Das außenpolitische Vokabular der Berliner Republik als Fährte einer Neuorientierung." *Politische Vierteljahresschrift* 48 (4): 650–79.

Hempel, Carl G. 1950. "Problems and Changes in the Empiricist Criterion of Meaning." *Revue internationale de philosophie* 4 (11): 41–63.

Hempel, Carl G. 1962. "Deductive-Nomological vs. Statistical Explanation." In *Minnesota Studies in the Philosophy of Science*, vol. 3, edited by H. Feigl and G. Maxwell, 98–169. Minneapolis: University of Minnesota Press.

Hendershot, Chris, and David Mutimer. 2018. "Critical Security Studies." In *The Oxford Handbook of International Security*, edited by Alexandra Gheciu and William C. Wohlforth. Oxford: Oxford University Press, https://doi.org/10.1093/oxfordhb/9780198777854.013.5.

Henry, Marsha. 2017. "Problematizing military masculinity, intersectionality and male vulnerability in feminist critical military studies." *Critical Military Studies* 3 (2): 182–99.

Herborth, Benjamin. 2010. "Rekonstruktive Forschungslogik." In *Handbuch der nternationalen Politik*, edited by Carlo Masala, Frank Sauer, and Andreas Wilhelm, 265–84. Wiesbaden: VS Verlag für Sozialwissenschaften.

Herborth, Benjamin. 2011. "Methodenstreit—Methodenzwang—Methoden-fetisch." *Zeitschrift für Internationale Beziehungen* 18 (2): 137–51.

Hermann, Charles F. 1990. "Changing Course: When Governments Choose to Redirect Foreign Policy." *International Studies Quarterly* 34 (1): 3–21.

Hermann, Margaret G. 2005. "Assessing Leadership Style: Trait Analysis." In *The Psychological Assessment of Political Leaders: With Profiles of Saddam Hussein and Bill Clinton*, edited by Jerrold M. Post, 178–214. Ann Arbor: University of Michigan Press.

Herschinger, Eva. 2011. *Constructing Global Enemies: Hegemony and Identity in International Discourses on Terrorism and Drug Prohibition*. London: Routledge.

Herschinger, Eva. 2012. "'Hell Is the Other': Conceptualising Hegemony and Identity through Discourse Theory." *Millennium: Journal of International Studies* 42 (1): 65–90.

Herzog, Benno. 2016. "Discourse Analysis as Immanent Critique: Possibilities and Limits of Normative Critique in Empirical Discourse Studies." *Discourse and Society* 27 (3): 278–92

Hewett, Ed A., and Victor H. Winston, eds. 1991. *Milestones in Glasnost and Perestroyka: Politics and People*. Washington, DC: Brookings Institution Press.

Higate, Paul, ed. 2003. *Military Masculinities: Identity and the State*. Westport, CT: Praeger.

Higate, Paul, and Ailsa Cameron. 2006. "Reflexivity and Researching the Military." *Armed Forces and Society* 32 (2): 219–33.

Hill, Christopher. 2016. *Foreign Policy in the Twenty-First Century*. 2nd ed. Basingstoke: Palgrave Macmillan.

Hilpert, Carolin. 2014. *Strategic Cultural Change and the Challenge for Security Policy: Germany and the Bundeswehr's Deployment to Afghanistan*. Basingstoke: Palgrave Macmillan.

Hindess, Barry. 1973. "Models and Masks: Empiricist Conceptions of the Conditions of Scientific Knowledge." *Economy and Society* 2 (2): 233–54.

Hobson, John M. 2012. *The Eurocentric Conception of World Politics: Western International Theory, 1760–2010*. Cambridge: Cambridge University Press.

Hoeffler, Anke. 2014. "Can International Interventions Secure the Peace?" *International Area Studies Review* 17 (1): 75–94.

Hoffman, Bruce. 2015. "A First Draft of the History of America's Ongoing Wars on Terrorism." *Studies in Conflict and Terrorism* 38 (1): 75–83.

Hoffman, Mark. 1987. "Critical Theory and the Inter-paradigm Debate." *Millennium: Journal of International Studies* 16 (2): 231–50.

Hoffmann, Stanley. 2001. "Why Don't They Like Us?" *American Prospect*, 19 December, https://prospect.org/features/like-us/.

Hofius, Maren, Jan Wilkens, Hannes Hansen-Magnusson, and Sassan Gholiagha. 2014. "Den Schleier lichten? Kritische Normenforschung, Freiheit und Gleichberechtigung im Kontext des 'Arabischen Frühlings.'" *Zeitschrift für Internationale Beziehungen* 21 (2): 85–105.

Hofmann, Stephanie C. 2019. "Beyond Culture and Power: The Role of Party Ideologies in German Foreign and Security Policy." *German Politics*, advance online article, https://doi.org/10.1080/09644008.2019.1611783.

Hofweber, Thomas. 2014. "Logic and Ontology." In *The Stanford Encyclopedia of Philosophy*, edited by Edward N. Zalta. Fall ed. Stanford: Stanford University, http://plato.stanford.edu/entries/logic-ontology/.

Holland, Jack. 2009. "From September 11th 2001 to 9-11: From Void to Crisis." *International Political Sociology* 3 (3): 275–92.

Holland, Jack. 2011. "Foreign Policy and Political Possibility." *European Journal of International Relations* 19 (1): 49–68.

Holland, Jack. 2012. *Selling the War on Terror: Foreign Policy Discourses after 9/11*. London: Routledge.

Holland, Lauren. 1999. "The U.S. Decision to Launch Operation Desert Storm: A Bureaucratic Politics Analysis." *Armed Forces and Society* 25 (2): 219–42.

Hollis, Martin, and Steve Smith. 2004. *Explaining and Understanding International Relations*. Oxford: Clarendon.

Holzapfel, Klaus-J. 1987. *Kürschners Volkshandbuch Deutscher Bundestag, 11. Wahlperiode*. 51st ed. Rheinbreitbach: NDV.

Holzapfel, Klaus-J. 1992. *Kürschners Volkshandbuch Deutscher Bundestag, 12. Wahlperiode*. 67th ed. Rheinbreitbach: NDV.

Holzapfel, Klaus-J. 1996. *Kürschners Volkshandbuch Deutscher Bundestag, 13. Wahlperiode*. 79th ed. Rheinbreitbach: NDV.

Holzapfel, Klaus-J. 1999. *Kürschners Volkshandbuch Deutscher Bundestag, 14. Wahlperiode*. 83rd ed. Rheinbreitbach: NDV.

Holzapfel, Klaus-J. 2003. *Kürschners Volkshandbuch Deutscher Bundestag, 15. Wahlperiode*. 95th ed. Rheinbreitbach: NDV.

Holzapfel, Klaus-J. 2009. *Kürschners Volkshandbuch Deutscher Bundestag, 16. Wahlperiode*. 112th ed. Rheinbreitbach: NDV.

Holzapfel, Klaus-J. 2012. *Kürschners Volkshandbuch Deutscher Bundestag, 17. Wahlperiode*. 126th ed. Rheinbreitbach: NDV.

Holzscheiter, Anna. 2014. "Between Communicative Interaction and Structures of Signification: Discourse Theory and Analysis in International Relations." *International Studies Perspectives* 15 (2): 142–62.

Hooper, Charlotte. 2001. *Manly States: Masculinities, International Relations, and Gender Politics*. New York: Columbia University Press.

Hopf, Ted. 1998. "The Promise of Constructivism in International Relations Theory." *International Security* 23 (1): 171–200.

Horkheimer, Max. 1937. "Traditionelle und kritische Theorie." *Zeitschrift für Sozialforschung* 6 (2): 245–94.

Hough, Daniel. 2000. "'Made in Eastern Germany': The PDS and the Articulation of Eastern German Interests." *German Politics* 9 (2): 125–48.

Houghton, David Patrick. 1996. "The Role of Analogical Reasoning in Novel Foreign-Policy Situations." *British Journal of Political Science* 26 (4): 523–52.

Howarth, David. 1995. "Discourse Theory." In *Theory and Methods in Political Sci-*

ence, edited by David Marsh and Gerry Stoker, 115–36. Basingstoke: Macmillan.

Howarth, David. 2000. *Discourse*. Buckingham, UK: Open University Press.

Howarth, David. 2005. "Applying Discourse Theory: The Method of Articulation." In *Discourse Theory in European Politics*, edited by David Howarth and Jacob Torfing, 316–46. Basingstoke: Palgrave Macmillan.

Howarth, David. 2008. "Ethos, Agonism and Populism: William Connolly and the Case for Radical Democracy." *British Journal of Politics and International Relations* 10 (2): 171–93.

Howarth, David. 2010. "Power, Discourse, and Policy: Articulating a Hegemony Approach to Critical Policy Studies." *Critical Policy Studies* 3 (3–4): 309–35.

Howarth, David. 2013. *Poststructuralism and After: Structure, Subjectivity and Power*. Basingstoke: Palgrave Macmillan.

Howarth, David, Aletta Norval, and Yannis Stavrakakis, eds. 2000. *Discourse Theory and Political Analysis: Identities, Hegemonies and Social Change*. Manchester: Manchester University Press.

Howarth, David, and Yannis Stavrakakis. 2000. "Introducing Discourse Theory and Political Analysis." In *Discourse Theory and Political Analysis: Identities, Hegemonies and Social Change*, edited by David Howarth, Aletta J. Norval and Yannis Stavrakakis, 1–23. Manchester: Manchester University Press.

Hudson, Heidi. 2012. "A Double-Edged Sword of Peace? Reflections on the Tension between Representation and Protection in Gendering Liberal Peacebuilding." *International Peacekeeping* 19 (4): 443–60.

Hudson, Valerie M., ed. 1997. *Culture and Foreign Policy*. Boulder, CO: Lynne Rienner.

Huggan, Graham. 2005. "(Not) Reading Orientalism." *Research in African Literatures* 36 (3): 124–36.

Hughes, Christopher W. 2004. *Japan's Re-emergence as a "Normal" Military Power*. Adelphi Papers no. 368–69. London: Routledge.

Hughes, Christopher W. 2009. *Japan's Remilitarisation*. Adelphi Papers no. 403 London: Routledge.

Hughes, Geraint. 2015. "Why Military Interventions Fail: An Historical Overview." *British Journal for Military History* 1 (2): 101–18.

Hurrelmann, Achim, Zuzana Krell-Laluhová, Frank Nullmeier, Steffen Schneider, and Achim Wiesner. 2009. "Why the Democratic Nation-State Is Still Legitimate: A Study of Media Discourses." *European Journal of Political Research* 48 (4): 483–515.

Hutchings, Kimberly. 1994. "Borderline Ethics: Feminist Morality and International Relations." *Paradigms* 8 (1): 23–35.

Hutchings, Kimberly. 2008. "Making Sense of Masculinity and War." *Men and Masculinities* 10 (4): 389–404.

Hutchings, Kimberly. 2018. "Pacifism Is Dirty: Towards an Ethico-Political Defence." *Critical Studies on Security* 6 (2): 176–92.

Hutchings, Kimberly. 2019. "From Just War Theory to Ethico-Political Pacifism." *Critical Studies on Security* 7 (3): 191–98.

Hyde-Price, Adrian. 2001. "Germany and the Kosovo War: Still a Civilian Power?" *German Politics* 10 (1): 19–34.

Hynek, Nik, and David Chandler. 2013. "No Emancipatory Alternative, No Critical Security Studies." *Critical Studies on Security* 1 (1): 46–63.

Hyslop, Alec. 2015. "Other Minds." In *Stanford Encyclopedia of Philosophy*, edited by Edward N. Zalta. Fall ed. Stanford, CA: Stanford University, http://plato.stanford.edu/entries/other-minds/.

Ilie, Cornelia. 2003. "Discourse and Metadiscourse in Parliamentary Debates." *Journal of Language and Politics* 2 (1): 71–92.

Inayatullah, Naeem. 2013a. "Playing on the Shores of an Imperial Pedagogy." *Critical Studies on Security* 1 (3): 355–57.

Inayatullah, Naeem. 2013b. "Pulling Threads: Intimate Systematicity in the Politics of Exile." *Security Dialogue* 44 (4): 331–45.

Inayatullah, Naeem. 2014. "Why Do Some People Think They Know What Is Good for Others?" In *Global Politics: A New Introduction*, edited by Jenny Edkins and Maja Zehfuss, 450–71. 2nd ed. London: Routledge.

Inayatullah, Naeem, and David L. Blaney. 2004. *International Relations and the Problem of Difference*. London: Routledge.

Independent Commission on Disarmament and Security Issues. 1982. *Common Security: A Blueprint for Survival*. New York: Simon and Schuster.

Independent International Commission on Kosovo. 2000. *The Kosovo Report: Conflict, International Response, Lessons Learned*. Oxford: Oxford University Press.

Innes, Alexandria J. 2010. "When the Threatened Become the Threat: The Construction of Asylum Seekers in British Media Narratives." *International Relations* 24 (4): 456–77.

International Institute of Strategic Studies. 2007. "Europe." *Military Balance* 107 (1): 93–186.

International Institute of Strategic Studies. 2018. "South Asia and Afghanistan." *Strategic Survey* 118 (1): 110–47.

International Institute of Strategic Studies. 2019. "Asia." *Strategic Survey* 119 (1): 68–133.

Ishiyama, John. 2014. "Replication, Research Transparency, and Journal Publications: Individualism, Community Models, and the Future of Replication Studies." *PS: Political Science and Politics* 47 (1): 78–83.

Ismayr, Wolfgang. 2013. *Der Deutsche Bundestag*. 3rd ed. Bonn: Bundeszentrale für politische Bildung.

Jaberg, Sabine. 2008. "Abschied von der Friedensnorm? Urteile des Bundesverfassungsgerichts, verteidigungspolitische Grundsatzdokumente und die friedenspolitische Substanz des Grundgesetzes." In *Berliner Friedenspolitik? Militärische Transformation—zivile Impulse—europäische Einbindung*, edited by Peter Schlotter, Wilhelm Nolte, and Renate Grasse, 83–106. Baden-Baden: Nomos.

Jabri, Vivienne. 2013. *The Postcolonial Subject: Claiming Politics / Governing Others in Late Modernity*. London: Routledge.

Jackson, Patrick Thaddeus. 2006. *Civilizing the Enemy: German Reconstruction and the Invention of the West*. Ann Arbor: University of Michigan Press.

Jackson, Patrick Thaddeus. 2010. *The Conduct of Inquiry in International Relations: Philosophy of Science and Its Implications for the Study of World Politics*. 1st ed. New York: Routledge.

Jackson, Patrick Thaddeus. 2015. "Fear of Relativism." *International Studies Perspectives* 16 (1): 13–22.

Jackson, Richard. 2005. *Writing the War on Terrorism: Language, Politics and Counter-Terrorism*. Manchester: Manchester University Press.

Jackson, Richard. 2007. "Constructing Enemies: 'Islamic Terrorism' in Political and Academic Discourse." *Government and Opposition* 42 (3): 394–426.

Jackson, Richard. 2018. "Pacifism: The Anatomy of a Subjugated Knowledge." *Critical Studies on Security* 6 (2): 160–75.

Jackson, Richard. 2019. "Pacifism and the Ethical Imagination in IR." *International Politics* 56 (2): 212–27.

Jacobs, Thomas. 2019. "Poststructuralist Discourse Theory as an Independent Paradigm for Studying Institutions: Towards a New Definition of 'Discursive Construction' in Institutional Analysis." *Contemporary Political Theory* 18 (3): 379–401.

Jacobsen, Hans-Adolf, and Hans-Jürgen Rautenberg. 1991. *Bundeswehr und europäische Sicherheitsordnung: Abschlussbericht der Unabhängigen Kommission für die künftigen Aufgaben der Bundeswehr*. Bonn: Bouvier Verlag.

Jäger, Siegfried. 2001. "Diskurs und Wissen." In *Handbuch Sozialwissenschaftliche Diskursanalyse*. Vol. 1, *Theorien und Methoden*, edited by Reiner Keller, Andreas Hirseland, Werner Schneider, and Willy Viehöver, 81–112. Wiesbaden: VS Verlag für Sozialwissenschaften.

Jansen, Carolin, and Jürgen Maier. 2012. "Die Causa zu Guttenberg im Spiegel der Printmedien." *Zeitschrift für Politikberatung* 5 (1): 3–12.

Jarausch, Konrad H. 1995. "Normalisierung oder Re-Nationalisierung? Zur Umdeutung der deutschen Vergangenheit." *Geschichte und Gesellschaft* 21 (4): 571–84.

Jarausch, Konrad H. 2010. "The Federal Republic at Sixty: Popular Myths, Actual Accomplishments and Competing Interpretations." *German Politics and Society* 28 (1): 10–29.

Jepperson, Ronald L., Alexander Wendt, and Peter J. Katzenstein. 1996. "Norms, Identity, and Culture in National Security." In *The Culture of National Security: Norms and Identity in World Politics*, edited by Peter J. Katzenstein, 33–75. New York: Columbia University Press.

Jervis, Robert. 1976. *Perception and Misperception in International Politics*. Princeton, NJ: Princeton University Press.

Jervis, Robert. 2006. "Understanding Beliefs." *Political Psychology* 27 (5): 641–63.

Johnson, Chalmers. 2004a. *Blowback: The Costs and Consequences of American Empire*. 2nd ed. New York: Holt.

Johnson, Chalmers. 2004b. *The Sorrows of Empire: Militarism, Secrecy, and the End of the Republic*. New York: Holt.

Jones, Karen. 2004. "Gender and Rationality." In *Oxford Handbook of Rationality*, edited by Alfred R. Mele and Piers Rawling, 301–19. Oxford: Oxford University Press.

Jørgensen, Marianne, and Louise Phillips. 2002. *Discourse Analysis as Theory and Method*. Thousand Oaks, CA: Sage.

Joseph, Jonathan. 2002. *Hegemony: A Realist Analysis*. New York: Routledge.

Junk, Julian, and Christopher Daase. 2013. "Germany." In *Strategic Cultures in Europe*, edited by Heiko Biehl, Bastian Giegerich, and Alexandra Jonas, 139–52. Wiesbaden: Springer VS.

Kaim, Markus. 2007. "Deutsche Auslandseinsätze in der Multilateralismusfalle?" In *Auslandseinsätze der Bundeswehr: Leitfragen, Entscheidungsspielräume und Lehren*, edited by Stefan Mair, 43–49. Berlin: Stiftung Wissenschaft und Politik.

Kaldor, Mary, Mary Martin, and Sabine Selchow. 2007. "Human Security: A New Strategic Narrative for Europe." *International Affairs* 83 (2): 273–88.

Karp, Regina. (2009. "Germany: A 'Normal' Global Actor? *German Politics* 18 (1): 12—35.

Katz, Elihu. 2009. "Media-Government-Public: Coalitions and Oppositions." *Communication Review* 12 (3): 199–204.

Katzenstein, Peter J. 2003. "Same War—Different Views: Germany, Japan, and Counterterrorism." *International Organization* 57 (4): 731–60.

Kaufmann, Chaim. 2004. "Threat Inflation and the Failure of the Marketplace of Ideas: The Selling of the Iraq War." *International Security* 29 (1): 5–48.

Keller, Reiner. 2013. *Doing Discourse Research: An Introduction for Social Scientists*. Thousand Oaks, CA: Sage.

Keohane, Robert O. 1984. *After Hegemony: Cooperation and Discord in the World Political Economy*. Princeton, NJ: Princeton University Press.

Kepplinger, Hans Mathias. 2009. *Politikvermittlung*. Wiesbaden: VS Verlag für Sozialwissenschaften.

Kessler, Oliver. 2010. "Risk." In *The Routledge Handbook of New Security Studies*, edited by J. Peter Burgess, 17–26. London: Routledge.

Kessler, Oliver, and Wouter Werner. 2008. "Extrajudicial Killing as Risk Management." *Security Dialogue* 39 (2–3): 289–308.

Khalilzad, Zalmay, and Daniel Byman. 2000. "Afghanistan: The Consolidation of a Rogue State." *Washington Quarterly* 23 (1): 65–78.

King, Gary, Robert O. Keohane, and Sidney Verba. 1994. *Designing Social Inquiry: Scientific Inference in Qualitative Research*. Princeton, NJ: Princeton Univerisity Press.

Kingdon, John W. 1995. *Agendas, Alternatives, and Public Choices*. 2nd ed. New York: HarperCollins.

Kinnvall, Catarina, Ian Manners, and Jennifer Mitzen. 2018. "Introduction to 2018 Special Issue of European Security: Ontological (in)Security in the European Union." *European Security* 27 (3): 249–65.

Kinnvall, Catarina, and Jennifer Mitzen. 2017. "An Introduction to the Special Issue: Ontological Securities in World Politics." *Cooperation and Conflict* 52 (1): 3–11.

Kinzer, Stephen. 1991. "War in the Gulf: Germany; Germans Are Told of Gulf-War Role." *New York Times*, 31 January 1991, http://www.nytimes.com/1991/01/31/world/war-in-the-gulf-germany-germans-are-told-of-gulf-war-role.html?pagewanted=print&src=pm.

Kirste, Knut, and Hanns W. Maull. 1996. "Zivilmacht und Rollentheorie." *Zeitschrift für Internationale Beziehungen* 3 (2): 283–312.

Klein, Bradley S. 1990. "How the West Was One: Representational Politics of NATO." *International Studies Quarterly* 34 (3): 311–25.

Kocka, Jürgen. 1988. "German History before Hitler: The Debate about the German Sonderweg." *Journal of Contemporary History* 23 (1): 3–16.

Koddenbrock, Kai. 2012. "Recipes for Intervention: Western Policy Papers Imagine the Congo." *International Peacekeeping* 19 (5): 549–64.

Koenig, Nicole. 2020. "Leading beyond Civilian Power: Germany's Role Reconception in European Crisis Management." *German Politics* 29 (1): 79–96.

Körber-Stiftung. 2014. *Einmischen oder zurückhalten? Ergebnisse einer repräsentativen Umfrage von TNS Infratest Politikforschung zur Sicht der Deutschen auf die Außenpolitik*. Hamburg: Körber-Stiftung. https://www.koerber-stiftung.de/umfrage-aussenpolitik-einmischen-oder-zuruechhalten-901.

Korte, Karl-Rudolf. 2007. "Bundeskanzleramt." In *Handbuch zur deutschen Außenpolitik*, edited by Gunther Hellmann, Reinhard Wolf, and Siegmar Schmidt, 201–9. Wiesbaden: VS Verlag für Sozialwissenschaften.

Koslowski, Rey, and Friedrich V. Kratochwil. 1994. "Understanding Change in International Politics: The Soviet Empire's Demise and the International System." *International Organization* 48 (2): 215–47.

Krasner, Stephen D., and Thomas Risse. 2014. "External Actors, State-Building, and Service Provision in Areas of Limited Statehood: Introduction." *Governance* 27 (4): 545–67.

Kratochwil, Friedrich. 2007a. "Of Communities, Gangs, Historicity and the Problem of Santa Claus: Replies to My Critics." *Journal of International Relations and Development* 10 (1): 57–78.

Kratochwil, Friedrich. 2007b. "Of False Promises and Good Bets: A Plea for a Pragmatic Approach to Theory Building (the Tartu Lecture)." *Journal of International Relations and Development* 10 (1): 1–15.

Krebs, Ronald R. 2015. *Narrative and the Making of US National Security*. Cambridge: Cambridge University Press.

Krebs, Ronald R. 2018. "The Politics of National Security." In *The Oxford Handbook of International Security*, edited by Alexandra Gheciu and William C. Wohlforth. Oxford: Oxford University Press, https://doi.org/10.1093/oxfordhb/9780198777854.013.42.

Krebs, Ronald R., and Patrick Thaddeus Jackson. 2007. "Twisting Tongues and

Twisting Arms: The Power of Political Rhetoric." *European Journal of International Relations* 13 (1): 35–66.

Krumer-Nevo, Michal, and Mirit Sidi. 2012. "Writing against Othering." *Qualitative Inquiry* 18 (4): 299–309.

Kühn, Florian P. 2013. "Post-Interventionist Zeitgeist: The Ambiguity of Security Policy." In *The Armed Forces: Towards a Post-Interventionist Era?*, edited by Gerhard Kümmel and Bastian Giegerich, 17–28. Wiesbaden: Springer VS.

Kundnani, Hans. 2012. "The Concept of 'Normality' in German Foreign Policy since Unification." *German Politics and Society* 30 (2): 38–58.

Kusow, Abdi Mohamed. 1994. "The Genesis of the Somali Civil War: A New Perspective." *Northeast African Studies* 1 (1): 31–46.

Kvale, Steinar. 1995. "The Social Construction of Validity." *Qualitative Inquiry* 1 (1): 19–40.

Laclau, Ernesto. 1989. "Politics and the Limits of Modernity." *Social Text* (21): 63–82.

Laclau, Ernesto. 1990a. "New Reflections on the Revolution of Our Time." In *New Reflections on the Revolution of Our Time*, edited by Ernesto Laclau, 3–85. London: Verso.

Laclau, Ernesto, ed. 1990b. *New Reflections on the Revolution of Our Time*. London: Verso.

Laclau, Ernesto. 1992. "Beyond Emancipation." *Development and Change* 23 (3): 121–37.

Laclau, Ernesto. 1994. "Introduction" *The Making of Political Identities*, edited by Ernesto Laclau, 1–10. London: Verso.

Laclau, Ernesto. 1995. "Subject of Politics, Politics of the Subject." *Differences: A Journal of Feminist Cultural Studies* 7 (1): 146–64.

Laclau, Ernesto, ed. 1996a. *Emancipation(S)*. London: Verso.

Laclau, Ernesto. 1996b. "The Time Is Out of Joint." In *Emancipation(S)*, edited by Ernesto Laclau, 66–83. London: Verso.

Laclau, Ernesto. 1996c. "Universalism, Particularism, and the Question of Identity." In *Emancipation(S)*, edited by Ernesto Laclau, 20–35. London: Verso.

Laclau, Ernesto. 1996d. "Why Do Empty Signifiers Matter to Politics?" In *Emancipation(S)*, edited by Ernesto Laclau, 36–46. London: Verso.

Laclau, Ernesto. 1997. "The Death and Resurrection of the Theory of Ideology." *MLN* 112 (3): 297–321.

Laclau, Ernesto. 2000. "Identity and Hegemony: The Role of Universality in the Constitution of Political Logics." In *Contingency, Hegemony, Universality: Contemporary Dialogues on the Left*, edited by Judith Butler, Ernesto Laclau, and Slavoj Žižek, 44–89. London: Verso.

Laclau, Ernesto. 2005a. *On Populist Reason*. New York: Verso.

Laclau, Ernesto. 2005b. "Populism: What's in a Name?" In *Populism and the Mirror of Democracy*, edited by Francisco Panizza, 32–49. London: Verso.

Laclau, Ernesto. 2006. "Ideology and Post-Marxism." *Journal of Political Ideologies* 11 (2): 103–14.

Laclau, Ernesto. 2012. "Reply." *Cultural Studies* 26 (2–3): 391–415.

Laclau, Ernesto. 2014a. "Articulation and the Limits of Metaphor." In *The Rhetorical Foundations of Society*, edited by Ernesto Laclau, 53–78. London: Verso.

Laclau, Ernesto. 2014b. "The Death and Resurrection of the Theory of Ideology." In *The Rhetorical Foundations of Society*, edited by Ernesto Laclau, 11–36. London: Verso.

Laclau, Ernesto. 2014c. "Ethics, Normativity and the Heteronomy of the Law." In *The Rhetorical Foundations of Society*, edited by Ernesto Laclau, 127–37. London: Verso.

Laclau, Ernesto. 2014d. "Introduction." *The Rhetorical Foundations of Society*, edited by Ernesto Laclau, 1–10. London: Verso.

Laclau, Ernesto. 2014e. "On the Names of God." In *The Rhetorical Foundations of Society*, edited by Ernesto Laclau, 37–51. London: Verso.

Laclau, Ernesto. 2014f. "The Politics of Rhetoric." In *The Rhetorical Foundations of Society*, edited by Ernesto Laclau, 79–99. London: Verso.

Laclau, Ernesto, ed. 2014g. *The Rhetorical Foundations of Society*. London: Verso.

Laclau, Ernesto, and Roy Bhaskar. 1998. "Discourse Theory vs. Critical Realism." *Journal of Critical Realism* 1 (2): 9–14.

Laclau, Ernesto, and Chantal Mouffe. 1987. "Post-Marxism without Apologies." *New Left Review* (166): 79–106.

Laclau, Ernesto, and Chantal Mouffe. 2001. *Hegemony and Socialist Strategy: Towards a Radical Democratic Politics*. 2nd ed. London: Verso.

Laclau, Ernesto, and Lilian Zac. 1994. "Minding the Gap: The Subject of Politics." In *The Making of Political Identities*, edited by Ernesto Laclau, 11–39. London: Verso.

Lagassé, Philippe, and Patrick A. Mello. 2018. "The Unintended Consequences of Parliamentary Involvement: Elite Collusion and Afghanistan Deployments in Canada and Germany." *British Journal of Politics and International Relations* 20 (1): 135–57.

Lambach, Daniel. 2015. "Repräsentationen von Territorialität und internationale Ungleichheit." *Zeitschrift für Außen- und Sicherheitspolitik* 8 (2): 243–69.

Lantis, Jeffrey. 2002a. "The Moral Imperative of Force: The Evolution of German Strategic Culture in Kosovo." *Comparative Strategy* 21 (21–46): 1.

Lantis, Jeffrey S. 2002b. "Strategic Culture and National Security Policy." *International Studies Review* 4 (3): 87–113.

Lanzara, Giovan Francesco. 1998. "Self-Destructive Processes in Institution Building and Some Modest Countervailing Mechanisms." *European Journal of Political Research* 33 (1): 1–39.

Laqueur, Walter, and Christopher Wall. 2018. *The Future of Terrorism: ISIS, Al-Qaeda, and the Alt-Right*. New York: St. Martin's.

Law, John. 2004. *After Method: Mess in Social Science Research*. New York: Routledge.

Law, John. 2009. "Seeing Like a Survey." *Cultural Sociology* 3 (2): 239–56.

Lebow, Richard Ned. 2008. "Identity and International Relations." *International Relations* 22 (4): 473–92.

Leek, Maria, and Viacheslav Morozov. 2018. "Identity beyond Othering: Crisis and the Politics of Decision in the EU's Involvement in Libya." *International Theory* 10 (1): 122–52.

Lefèvre, Raphaël. 2018. "The Resurgence of Al-Qaeda in the Islamic Maghrib." *Journal of North African Studies* 23 (1–2): 278–81.

Leffler, Melvyn P. 2003. "9/11 and the Past and Future of American Foreign Policy." *International Affairs* 79 (5): 1045–63.

Legro, Jeffrey W., and Andrew Moravcsik. 1999. "Is Anybody Still a Realist?" *International Security* 24 (2): 5–55.

Legro, Jeffrey W., and Andrew Moravcsik. 2000. "Brother, Can You Spare a Paradigm? (or Was Anybody Ever a Realist?)." *International Security* 25 (1): 184–93.

Lehmbruch, Gerhard. 1998. *Parteienwettbewerb im Bundesstaat: Regelsysteme und Spannungslagen im Institutionengefüge der Bundesrepublik Deutschland.* 2nd ed. Opladen: Westdeutscher Verlag.

Leibfried, Stephan, Evelyne Huber, Matthew Lange, Jonah H. Levy, Frank Nullmeier, and John D. Stephens, eds. 2015. *The Oxford Handbook of Transformations of the State.* Oxford: Oxford University Press.

Leithner, Anika. 2009. *Shaping German Foreign Policy: History, Memory, and National Interest.* Boulder, CO: FirstForumPress.

Leonhard, Nina. 2019. "Towards a New German Military Identity? Change and Continuity of Military Representations of Self and Other(s) in Germany." *Critical Military Studies* 5 (4): 304–21.

Lepsius, M. Rainer. 1964. "Kritik als Beruf, zur Soziologie der Intellektuellen." *Kölner Zeitschrift für Soziologie und Sozialpsychologie* 16 (1): 75–91.

Levitsky, Steven, and Daniel Ziblatt. 2018. *How Democracies Die.* New York: Crown.

Lieven, Anatol. 1994. *The Baltic Revolution: Estonia, Latvia, Lithuania and the Path to Independence.* New Haven, CT: Yale University Press.

Linz, Juan J., and Alfred Stepan. 1992. "Political Identities and Electoral Sequences: Spain, the Soviet Union, and Yugoslavia." *Dædalus* 121 (2): 123–39.

Lockyer, Adam. 2010. "The Dynamics of Warfare in Civil War." *Civil Wars* 12 (1–2): 91–116.

Longerich, Peter. 2006. *"Davon haben wir nichts gewusst!" Die Deutschen und die Judenverfolgung 1933–1945.* Bonn: Bundeszentrale für politische Bildung.

Longhurst, Kerry. 2004. *Germany and the Use of Force.* Manchester: Manchester University Press.

Lund, Michael. 2002. "Preventing Violent Intrastate Conflicts: Learning Lessons from Experience." In *Searching for Peace in Europe and Eurasia: An Overview of Conflict Prevention and Peacebuilding Activities,* edited by Paul van Tongeren, Hans van de Veen, and Juliette Verhoeven, 99–119. London: Lynne Rienner.

Lundborg, Tom. 2012. *Politics of the Event: Time, Movement, Becoming.* London: Routledge.

Macgilchrist, Felicitas, Johanna Ahlrichs, Patrick Mielke, and Roman Richtera. 2017. "Memory Practices and Colonial Discourse: On Text Trajectories and Lines of Flight." *Critical Discourse Studies* 14 (4): 341–61.

Mac Ginty, Roger. 2012. "Against Stabilization." *Stability: International Journal of Security and Development* 1 (1): 20–30.

Mader, Matthias. 2017. *Öffentliche Meinung zu Auslandseinsätzen der Bundeswehr: Zwischen Antimilitarismus und transatlantischer Orientierung*. Wiesbaden: Springer VS.

Mahoney, James. 2010. "After KKV: The New Methodology of Qualitative Research." *World Politics* 62 (1): 120–47.

Malici, Akan. 2006. "Germans as Venutians: The Culture of German Foreign Policy Behavior." *Foreign Policy Analysis* 2 (1): 37–62.

Mälksoo, Maria. 2015. "'Memory Must Be Defended': Beyond the Politics of Mnemonical Security." *Security Dialogue* 46 (3): 221–37.

Mamdani, Mahmood. 2002. "Good Muslim, Bad Muslim: A Political Perspective on Culture and Terrorism." *American Anthropologist* 104 (3): 766–75.

Mamdani, Mahmood. 2005. *Good Muslim, Bad Muslim: America, the Cold War, and the Roots of Terror*. New York: Doubleday.

Mandelbaum, Michael. 1992. "Coup de Grace: The End of the Soviet Union." *Foreign Affairs* 71 (1): 164–83.

Mansfield, Edward D., and Jack Snyder. 1995. "Democratization and the Danger of War." *International Security* 20 (1): 5–38.

Mansfield, Edward D., and Jack Snyder. 2002. "Democratic Transitions, Institutional Strength, and War." *International Organization* 56 (2): 297–337.

Mansfield, Edward D., and Jack Snyder. 2009. "Pathways to War in Democratic Transitions." *International Organization* 63 (2): 381–90.

Mansfield, Nick. 2000. *Subjectivity: Theories of the Self from Freud to Haraway*. St. Leonards: Allen and Unwin.

March, James G., and Johan P. Olsen. 1998. "The Institutional Dynamics of International Political Orders." *International Organization* 52 (4): 943–69.

Marchart, Oliver. 2007. *Post-Foundational Political Thought: Political Difference in Nancy, Lefort, Badiou and Laclau*. Edinburgh: Edinburgh University Press.

Markoff, John. 2011. "A Moving Target: Democracy." *European Journal of Sociology* 52 (2): 239–76.

Marschall, Stefan. 2011. *Das politische System Deutschlands*. 2nd ed. Konstanz: UVK.

Marsh, David. 2015. "Two cheers for Boswell and Corbett." *Critical Policy Studies* 9 (2):230–33.

Marsh, David, and Paul Furlong. 2002. "A Skin Not a Sweater: Ontology and Epistemology in Political Science." In *Theory and Methods in Political Science*, edited by David Marsh and Gerry Stoker, 17–41. London: Palgrave Macmillan.

Martinsen, Kaare Dahl. 2010. "National Interests in German Security White Books." *National Identities* 12 (2): 161–80.

Martinsen, Kaare Dahl. 2013. "Totgeschwiegen? Deutschland und die Gefallenen des Afghanistan-Einsatzes." *Aus Politik und Zeitgeschichte* 63 (44): 17–23.

Martinson, Jeffrey D. 2012. "Rediscovering Historical Memory: German Foreign

Military Intervention Decision Making through the Second Lebanon War." *Foreign Policy Analysis* 8 (4): 389–407.

Marttila, Tomas, and Vincent Gengnagel. 2015. "Post-Foundational Discourse Analysis and the Impasses of Critical Inquiry." *Zeitschrift für Diskursforschung* 3 (1): 52–69.

Mason, Michael, and Mark Zeitoun. 2013. "Questioning Environmental Security." *Geographical Journal* 179 (4): 294–97.

Masters, Cristina. 2009. "Femina Sacra: The 'War on/of Terror,' Women and the Feminine." *Security Dialogue* 40 (1): 29–49.

Matin, Kamran. 2012. "Redeeming the Universal: Postcolonialism and the Inner Life of Eurocentrism." *European Journal of International Relations* 19 (2): 353–77.

Maull, Hanns W. 1990. "Germany and Japan: The New Civilian Powers." *Foreign Affairs* 69 (5): 91–106.

Maull, Hanns W. 2000. "Germany and the Use of Force: Still a Civilian Power?" *Survival* 42 (2): 56–80.

Maull, Hanns W. 2006. "Die prekäre Kontinuität: Deutsche Außenpolitik zwischen Pfadabhängigkeit und Anpassungsdruck." In *Regieren in der Bundesrepublik Deutschland: Innen- und Außenpolitik seit 1949*, edited by Manfred G. Schmidt and Reimut Zohlnhöfer, 421–45. Wiesbaden: VS Verlag für Sozialwissenschaften.

Maull, Hanns W. 2014. "'Zivilmacht': Ursprünge und Entwicklungspfade eines umstrittenen Konzeptes." In *Deutsche Außenpolitik und internationale Führung: Ressourcen, Praktiken und Politiken in einer veränderten Europäischen Union*, edited by Sebastian Harnisch and Joachim Schild, 121–47. Baden-Baden: Nomos.

Maull, Hanns W. 2018. "Reflective, Hegemonic, Geo-Economic, Civilian...? The Puzzle of German Power." *German Politics* 27 (4): 460–78.

Mayer, Peter. 2003. "Die Epistemologie der Internationalen Beziehungen: Anmerkungen zum Stand der 'Dritten Debatte.'" In *Die neuen Internationalen Beziehungen: Forschungsstand und Perspektiven in Deutschland*, edited by Gunther Hellmann, Klaus Dieter Wolf, and Michael Zürn, 47–97. Baden-Baden: Nomos.

Mays, Nicholas, and Catherine Pope. 1995. "Rigour and Qualitative Research." *BMJ* 311: 109–12.

McCourt, David M. 2012. "The Roles States Play: A Meadian Interactionist Approach." *Journal of International Relations and Development* 15 (3): 370–92.

McDonald, Matt. 2012. "The Failed Securitization of Climate Change in Australia." *Australian Journal of Political Science* 47 (4): 579–92.

McMahan, Jeff. 2009. *Killing in War*. Oxford: Oxford University Press.

Meiers, Franz-Josef. 2011. "Der wehrverfassungsrechtliche Parlamentsvorbehalt und die Verteidigung der Sicherheit Deutschlands am Hindukusch, 2001–2011." *Zeitschrift für Außen- und Sicherheitspolitik* 4 (S1): 87–113.

Meiers, Franz-Josef. 2012. "Zivilmacht als Willensfanatiker: Die libysche

Deutschstunde." In *Die Außenpolitik der Bundesrepublik Deutschland: Anpruch, Realität, Perspektiven*, edited by Reinhard Meier-Walser and Alexander Wolf, 161–73. Munich: Hanns-Seidel-Stiftung.

Meuche-Mäker, Meinhard. 2005. *Die PDS im Westen 1990–2005: Schlussfolgerungen für eine neue Linke*. Berlin: Karl Dietz Verlag.

Meusch, Andreas. 2008. "Das Einsatzführungskommando der Bundeswehr." *Wehrtechnik* 40 (5): 24–31.

Meyer, Berthold. 2006. "Die Parlamentsarmee—Zu schön, um wahr zu sein?" In *Armee in der Demokratie: Zum Verhältnis von zivilen und militärischen Prinzipien*, edited by Ulrich vom Hagen, 51–72. Wiesbaden: VS Verlag für sozialwissenschaften.

Mignolo, Walter D. 2007. "Delinking: The Rhetoric of Modernity, the Logic of Coloniality and the Grammar of De-Coloniality." *Cultural Studies* 21 (2–3): 449–514.

Mignolo, Walter D., and Arturo Escobar, eds. 2010. *Globalization and the Decolonial Option*. London: Routledge.

Millar, Katharine M. 2018. "What Do We Do Now? Examining Civilian Masculinity/ies in Contemporary Liberal Civil-Military Relations." *Review of International Studies* 45 (2): 239–59.

Miller, Alexander. 2014. "Realism." In *The Stanford Encyclopedia of Philosophy*, edited by Edward N. Zalta. Winter ed. Stanford: Stanford University, http://plato.stanford.edu/entries/realism/.

Miller, Benjamin. 2010. "Explaining Changes in U.S. Grand Strategy: 9/11, the Rise of Offensive Liberalism, and the War in Iraq." *Security Studies* 19 (1): 26–65.

Miller, Tina. 2005. *Making Sense of Motherhood: A Narrative Approach*. Cambridge: Cambridge University Press.

Milliken, Jennifer. 1999. "The Study of Discourse in International Relations: A Critique of Research and Methods." *European Journal of International Relations* 5 (2): 225–54.

Mills, Charles W. 2014. *The Racial Contract*. Ithaca, NY: Cornell University Press.

Mintz, Alex, and Steven B. Redd. 2003. "Framing Effects in International Relations." *Synthese* 135 (2): 193–213.

Miskimmon, A. 2012. "German Foreign Policy and the Libya Crisis." *German Politics* 21 (4): 392–410.

Miskovic, Natasa, Harald Fischer-Tiné, and Nada Boskovska, eds. 2014. *The Nonaligned Movement and the Cold War: Delhi—Bandung—Belgrade*. London: Routledge.

Mitzen, Jennifer. 2006. "Ontological Security in World Politics: State Identity and the Security Dilemma." *European Journal of International Relations* 12 (3): 341–70.

Mitzen, Jennifer. 2015. "Illusion or Intention? Talking Grand Strategy into Existence." *Security Studies* 24 (1): 61–94.

Moghadam, Assaf. 2012. "Failure and Disengagement in the Red Army Faction." *Studies in Conflict and Terrorism* 35 (2): 156–81.

Molt, Matthias. 2007. "Von der Wehrmacht zur Bundeswehr: Personelle Kontinuität und Diskontinuität beim Aufbau der Deutschen Streitkräfte 1955–1966." PhD diss., Universität Heidelberg, https://archiv.ub.uni-heidelberg.de/volltextserver/8935/.

Monteiro, Nuno P., and Keven G. Ruby. 2009. "IR and the False Promise of Philosophical Foundations." *International Theory* 1 (1): 15.

Moravcsik, Andrew. 1997. "Taking Preferences Seriously: A Liberal Theory of International Politics." *International Organization* 51 (4): 513–53.

Moreau, Patrick. 2013. "Arbeit und soziale Gerechtigkeit—die Wahlalternative (WASG)." In *Handbuch der deutschen Parteien*, edited by Frank Decker and Viola Neu, 147–52. Wiesbaden: Springer VS.

Moses, Jonathon W., and Torbjørn L. Knutsen. 2007. *Ways of Knowing: Competing Methodologies in Social and Political Research*. Basingstoke: Palgrave Macmillan.

Mouffe, Chantal. 2005a. *On the Political*. London: Routledge.

Mouffe, Chantal. 2005b. *The Return of the Political*. 2nd ed. London: Verso.

Mueller, John. 2005. "Simplicity and Spook: Terrorism and the Dynamics of Threat Exaggeration." *International Studies Perspectives* 6 (2): 208–34.

Mueller, John, and Mark G. Stewart. 2018. "Terrorism and Bathtubs: Comparing and Assessing the Risks." *Terrorism and Political Violence*, advance online article, https://doi.org/10.1080/09546553.2018.1530662.

Müller, Harald. 2000. "Do Not Send the Marines! Plädoyer für die Einrichtung eines Ministeriums für Krisenprävention." *E+Z: Entwicklung und Zusammenarbeit* 41 (9): 232–33.

Müller, Harald. 2004a. "The Antinomy of Democratic Peace." *International Politics* 41 (4): 494–520.

Müller, Harald. 2004b. "Arguing, Bargaining and All That: Communicative Action, Rationalist Theory and the Logic of Appropriateness in International Relations." *European Journal of International Relations* 10 (3): 395–435.

Müller, Harald, and Jonas Wolff. 2011. "Demokratischer Krieg am Hindukusch? Eine kritische Analyse der Bundestagsdebatten zur deutschen Afghanistanpolitik 2001–2011." *Zeitschrift für Außen- und Sicherheitspolitik* 4 (S1): 197–221.

Müller, Ulrike Anne. 2011. "Far Away So Close: Race, Whiteness, and German Identity." *Identities* 18 (6): 620–45.

Münch, Richard. 2014. *Academic Capitalism: Universities in the Global Struggle for Excellence*. London: Routledge.

Muppidi, Himadeep. 2012. *The Colonial Signs of International Relations*. London: Hurst.

Muppidi, Himadeep. 2013. "On the Politics of Exile." *Security Dialogue* 44 (4): 299–313.

Muppidi, Himadeep. 2018. "Coloring the Global: Race, Colonialism and Internationalism." In *The Sage Handbook of the History, Philosophy and Sociology of International Relations*, edited by Andreas Gofas, Inanna Hamati-Ataya, and Nicholas Onuf, 46–59. London: Sage.

Murtazashvili, Jennifer Brick. 2016. "Afghanistan: A Vicious Cycle of State Failure." *Governance* 29 (2): 163–66.

Mustapha, Jennifer. 2013. "Ontological Theorizations in Critical Security Studies: Making the Case for a (Modified) Post-Structuralist Approach." *Critical Studies on Security* 1 (1): 64–82.

Nabers, Dirk. 2009. "Filling the Void of Meaning: Identity Construction in U.S. Foreign Policy after September 11, 2001." *Foreign Policy Analysis* 5 (2): 191–214.

Nabers, Dirk. 2015. *A Poststructuralist Discourse Theory of Global Politics*. Basingstoke: Palgrave Macmillan.

Nabers, Dirk. 2017. "Crisis as Dislocation in Global Politics." *Politics* 37 (4): 418–31.

Nabers, Dirk. 2018. "Towards International Relations Beyond the Mind." *Journal of International Political Theory* 16 (1): 89–105.

Nabers, Dirk. 2019. "Discursive Dislocation: Toward a Poststructuralist Theory of Crisis in Global Politics." *New Political Science* 41 (2): 263–78.

Nabers, Dirk, and Frank A. Stengel. 2019a. "International/Global Political Sociology." In *The Oxford Research Encyclopedia of International Studies*, edited by Renée Marlin-Bennett. Oxford: Oxford University Press, https://doi.org/10.1093/acrefore/9780190846626.013.371.

Nabers, Dirk, and Frank A. Stengel. 2019b. "Sedimented Practices in Trump's Foreign Policy." In *Populism and World Politics: Exploring Inter- and Transnational Dimensions*, edited by Frank A. Stengel, David B. MacDonald, and Dirk Nabers, 103–35. Basingtoke: Palgrave Macmillan.

Nachtwei, Winfried. 2009. "Jetzt ist endlich Krach in der Bude." *Internationale Politik* 64 (5): 1–4.

Nagel, Thomas. 1974. "What Is It Like to Be a Bat?" *Philosophical Review* 83 (4): 435–50.

NATO. 1999. *The Alliance's Strategic Concept*. Brussels: NATO, https://www.nato.int/cps/en/natolive/official_texts_27433.htm.

NATO. 2010. *Active Engagement, Modern Defence: Strategic Concept for the Defence and Security of the Members of the North Atlantic Treaty Organization*. Brussels: NATO, https://www.nato.int/strategic-concept/pdf/Strat_Concept_web_en.pdf.

Nayak, Anoop. 2007. "Critical Whiteness Studies." *Sociology Compass* 1 (2): 737–55.

Neocleous, Mark. 2011. "O Effeminacy! Effeminacy!": War, Masculinity and the Myth of Liberal Peace." *European Journal of International Relations* 19 (1): 93–113.

Nesbitt-Larking, Paul, and James W. McAuley. 2017. "Securitisation through Re-enchantment: The Strategic Uses of Myth and Memory." *Postcolonial Studies* 20 (3): 317–32.

Neu, Viola. 2013. "Die Linke." In *Handbuch der deutschen Parteien*, edited by Frank Decker and Viola Neu, 316–31. Wiesbaden: Springer VS.

Neumann, Cecilie Basberg, and Iver B Neumann. 2017. *Power, Culture and Situated Research Methodology: Autobiography, Field, Text*. Basingstoke: Palgrave Macmillan.

Neumann, Iver B. 1996. "Self and Other in International Relations." *European Journal of International Relations* 2 (2): 139–74.

Neumann, Iver B. 1999. *Uses of the Other: "The East" in European Identity Formation*. Minneapolis: University of Minnesota Press.

Neumann, Iver B. 2010. "Autobiography, Ontology, Autoethnology." *Review of International Studies* 36 (4): 1051–55.

Neureuther, Jörg. 2012. "Vernetzung in Sicherheitspolitik und militärischer Operationsführung: Versuch einer Bilanz." *Zeitschrift für Außen- und Sicherheitspolitik* 5 (1): 85–100.

Nicholson, Michael. 1996. "The Continued Significance of Positivism?" In *International Theory: Positivism and Beyond*, edited by Steve Smith, Ken Booth, and Marysia Zalewski, 128–45. Cambridge: Cambridge University Press.

Niclauß, Karlheinz. 1987. "Repräsentative und plebiszitäre Elemente der Kanzlerdemokratie." *Vierteljahrshefte für Zeitgeschichte* 35 (2): 217–45.

Niedermayer, Oskar. 2013. "Die Entwicklung des bundesdeutschen Parteiensystems." In *Handbuch der deutschen Parteien*, edited by Frank Decker and Viola Neu, 111–32. Wiesbaden: Springer VS.

Niemann, Holger, and Henrik Schillinger. 2017. "Contestation 'All the Way Down'? The Grammar of Contestation in Norm Research." *Review of International Studies* 43 (1): 29–49.

Nilsson, Desirée. 2008. "Partial Peace: Rebel Groups inside and outside of Civil War Settlements." *Journal of Peace Research* 45 (4): 479–95.

Noetzel, Timo. 2011. "The German Politics of War: Kunduz and the War in Afghanistan." *International Affairs* 87 (2): 397–417.

Nonhoff, Martin. 2001. "Soziale Marktwirtschaft—ein leerer Signifikant? Überlegungen im Anschluss an die Diskurstheorie Ernesto Laclaus." In *Diskursanalyse: Theorie, Methoden, Anwendungen*, edited by Johannes Angermüller, Katharina Bunzmann, and Martin Nonhoff, 193–208. Hamburg: Argument Verlag.

Nonhoff, Martin. 2006. *Politischer Diskurs und Hegemonie: Das Projekt "Soziale Marktwirtschaft."* Bielefeld: Transcript Verlag.

Nonhoff, Martin. 2007a. "Diskurs, radikale Demokratie, Hegemonie—Einleitung." In *Diskurs—radikale Demokratie—Hegemonie: Zum politischen Denken von Ernesto Laclau und Chantal Mouffe*, edited by Martin Nonhoff, 7–23. Bielefeld: Transcript Verlag.

Nonhoff, Martin. 2007b. "Politische Diskursanalyse als Hegemonieanalyse." In *Diskurs—radikale Demokratie—Hegemonie: Zum politischen Denken von Ernesto Laclau und Chantal Mouffe*, edited by Martin Nonhoff, 173–94. Bielefeld: Transcript.

Nonhoff, Martin. 2008. "Hegemonieanalyse: Theorie, Methode und Forschungspraxis." In *Handbuch Sozialwissenschaftliche Diskursanalyse, vol. 2, Forschungspraxis*, edited by Reiner Keller, Andreas Hirseland, Werner Schneider, and Willy Viehöver, 299–331. Wiesbaden: VS Verlag für Sozialwissenschaften.

Nonhoff, Martin. 2011. "Konstruktivistisch-pragmatische Methodik: Ein Plädoyer für die Diskursanalyse." *Zeitschrift für Internationale Beziehungen* 18 (2): 91–107.

Nonhoff, Martin. 2012. "Soziale Marktwirtschaft für Europa und die ganze Welt! Zur Legitimation ökonomischer Hegemonie in den Reden Angela Merkels." In *Der Aufstieg der Legitimitätspolitik: Rechtfertigung und Kritik politischökonomischer Ordnungen*, edited by Anna Geis, Frank Nullmeier, and Christopher Daase, 262-82. Baden-Baden: Nomos.

Nonhoff, Martin. 2017a. "Antagonismus und Antagonismen—hegemonietheoretische Aufklärung." In *Ordnungen des Politischen: Einsätze und Wirkungen der Hegemonietheorie Ernesto Laclaus*, edited by Oliver Marchart, 81-102. Wiesbaden: Springer VS.

Nonhoff, Martin. 2017b. "Discourse Analysis as Critique." *Palgrave Communications* 3: 17074.

Nonhoff, Martin. 2019. "Hegemony Analysis: Theory, Methodology and Research Practice." In *Discourse, Culture and Organization: Inquiries into Relational Structures of Power*, edited by Tomas Marttila, 63-104. Cham: Palgrave Macmillan.

Nonhoff, Martin, and Jennifer Gronau. 2012. "Die Freiheit des Subjekts im Diskurs." In *Diskurs—Macht—Subjekt*, edited by Reiner Keller, Werner Schneider, and Willy Viehöver, 109-30. Wiesbaden: VS Verlag für Sozialwissenschaften.

Nonhoff, Martin, and Frank A. Stengel. 2014. "Poststrukturalistische Diskurstheorie und Außenpolitikanalyse: Wie lässt sich Deutschlands wankelmütige Außenpolitik zwischen Afghanistan und Irak verstehen?" In *Diskursforschung in den Internationalen Beziehungen*, edited by Eva Herschinger and Judith Renner, 39-74. Baden-Baden: Nomos.

Nørgaard, Asbjørn S. 2008. "Political Science: Witchcraft or Craftsmanship? Standards for Good Research." *World Political Science Review* 4 (1): art. 5.

Norval, Aletta J. 2004. "Hegemony after Deconstruction: The Consequences of Undecidability." *Journal of Political Ideologies* 9 (2): 139-57.

Nuñez-Mietz, Fernando G. 2018. "Legalization and the Legitimation of the Use of Force: Revisiting Kosovo." *International Organization* 72 (3): 725-57.

Nymalm, Nicola. 2013. "The End of the 'Liberal Theory of History'? Dissecting the US Congress' Discourse on China's Currency Policy." *International Political Sociology* 7 (4): 388-405.

O'Driscoll, Cian. 2018. "The Irony of Just War." *Ethics and International Affairs* 32 (2): 227-36.

Olsen, John Andreas. 2013. *Strategic Air Power in Desert Storm*. London: Routledge.

Oppermann, Kai, and Alexander Höse. 2011. "Die innenpolitischen Restriktionen deutscher Außenpolitik." In *Deutsche Außenpolitik*, edited by Thomas Jäger, Alexander Höse, and Kai Oppermann, 44-76. Wiesbaden: VS Verlag für Sozialwissenschaften.

Orford, Anne. 1999. "Muscular Humanitarianism: Reading the Narratives of the New Interventionism." *European Journal of International Law* 10 (4): 679-711.

Ostermann, Falk. 2018. *Security, Defense Discourse and Identity in NATO and Europe: How France Changed Foreign Policy*. 1st ed. London: Routledge.

Ó Tuathail, Gearóid. 1994. "(Dis)placing Geopolitics: Writing on the Maps of Global Politics." *Environment and Planning D: Society and Space* 12 (5): 525-46.

Ó Tuathail, Gearóid. 1996. *Critical Geopolitics: The Politics of Writing Global Space.* Minneapolis: University of Minnesota Press.

Owens, Patricia. 2003. "Accidents Don't Just Happen: The Liberal Politics of High-Technology 'Humanitarian' War." *Millennium: Journal of International Studies* 32 (3): 595–616.

Panizza, Francisco, and Romina Miorelli. 2013. "Taking Discourse Seriously: Discursive Institutionalism and Post-Structuralist Discourse Theory." *Political Studies* 61 (2): 301–18.

Panke, Diana, and Ulrich Petersohn. 2012. "Why International Norms Disappear Sometimes." *European Journal of International Relations* 18 (4): 719–42.

Paris, Roland. 2014. "The 'Responsibility to Protect' and the Structural Problems of Preventive Humanitarian Intervention." *International Peacekeeping* 21 (5): 569–603.

Pashler, Harold, and Eric-Jan Wagenmakers. 2012. "Editors' Introduction to the Special Section on Replicability in Psychological Science: A Crisis of Confidence?" *Perspectives on Psychological Science* 7 (6): 528–30.

Patman, Robert G. 2015. "The Roots of Strategic Failure: The Somalia Syndrome and Al Qaeda's Path to 9/11." *International Politics* 52 (1): 89–109.

Patomäki, Heikki, and Colin Wight. 2000. "After Postpositivism? The Promises of Critical Realism." *International Studies Quarterly* 44 (2): 213–37.

Patzelt, Werner J. 1997. "German MPs and Their Roles." *Journal of Legislative Studies* 3 (1): 55–78.

Perger, Tilmann. 2002. "Ehrenschutz von Soldaten in Deutschland und anderen Staaten." PhD diss., Universität der Bundeswehr München, https://d-nb.info/972736786/34.

Perraudin, Michael, and Jürgen Zimmerer, eds. 2010. *German Colonialism and National Identity*. London: Routledge.

Peters, Dirk, Wolfgang Wagner, and Cosima Glahn. 2014. "Parliamentary Control of CSDP: The Case of the EU's Fight against Piracy off the Somali Coast." *European Security* 23 (4): 430–48.

Peterson, V. Spike. 2010. "Gendered Identities, Ideologies, and Practices in the Context of War and Militarism." In *Gender, War, and Militarism: Feminist Perspectives,* edited by Laura Sjoberg and Sandra Via, 17–29. Santa Barbara, CA: ABC-CLIO.

Peterson, V. Spike, and Anne Sisson Runyan. 1993. *Global Gender Issues.* Boulder, CO: Westview.

Pettersson, Mikael. 2011. "Depictive Traces: On the Phenomenology of Photography." *Journal of Aesthetics and Art Criticism* 69 (2): 185–96.

Pfeifer, Hanna, and Kilian Spandler. 2014. "The Responsibility to be Responsible: Über Außenpolitik und Verantwortung. *Wissenschaft und Frieden* (4): 36–39.

Pflieger, Klaus. 2011. *Die Rote Armee Fraktion.* 3rd ed. Baden-Baden: Nomos.

Philippi, Nina. 1997. *Bundeswehr-Auslandseinsätze als außen- und sicherheitspolitisches Problem des geeinten Deutschland.* Frankfurt am Main: Peter Lang.

Pluchinsky, Dennis A. 1993. "Germany's Red Army Faction: An Obituary." *Studies in Conflict and Terrorism* 16 (2): 135–57.

Popper, Karl. 2005. *The Logic of Scientific Discovery*. London: Routledge.

Porter, Patrick. 2018. "Why America's Grand Strategy Has Not Changed: Power, Habit, and the U.S. Foreign Policy Establishment." *International Security* 42 (4): 9–46.

Posen, Barry R. 1986. *The Sources of Military Doctrine: France, Britain, and Germany between the World Wars*. Ithaca, NY: Cornell University Press.

Probst, Lothar. 2013. "Bündnis 90 / die Grünen (Grüne)." In *Handbuch der deutschen Parteien*, edited by Frank Decker and Viola Neu, 166–79. Wiesbaden: Springer VS.

Prozorov, Sergei. 2011. "The Other as Past and Present: Beyond the Logic of 'Temporal Othering' in IR Theory." *Review of International Studies* 37 (3): 1273–93.

Prunier, Gérard. 1996. "Somalia: Civil War, Intervention and Withdrawal (1990–1995)." *Refugee Survey Quarterly* 15 (1): 35–85.

Puetter, U., and Antje Wiener. 2007. "Accommodating Normative Divergence in European Foreign Policy Co-ordination: The Example of the Iraq Crisis." *JCMS: Journal of Common Market Studies* 45 (5): 1065–88.

Quijano, Aníbal. 2007. "Coloniality and Modernity/Rationality." *Cultural Studies* 21 (2–3): 168–78.

Raap, Christian. 1996. "Änderungen im Wehrrecht." *Neue Zeitschrift für Verwaltungsrecht* 15 (5): 457.

Raap, Christian. 2002. "Änderungen im Wehrrecht." *Neue Zeitschrift für Verwaltungsrecht* 21 (8): 959–61.

Ramberg, Bjørn, and Kristin Gjesdal. 2014. "Hermeneutics." In *The Stanford Encyclopedia of Philosophy*, edited by Edward N. Zalta. Winter ed. Stanford: Stanford University, http://plato.stanford.edu/archives/spr2009/entries/rorty/.

Ramsbotham, Oliver, Hugh Miall, and Tom Woodhouse. 2011. *Contemporary Conflict Resolution*. 3rd ed. Oxford: Polity.

Rancière, Jacques. 1992. "Politics, Identification, and Subjectivization." *October* 61: 58–64.

Rau, Markus. 2006. "Auslandseinsatz der Bundeswehr: Was bringt das Parlamentsbeteiligungsgesetz?" *Archiv des Völkerrechts* 44 (1): 93–113.

Ray, James Lee. 2013. *American Foreign Policy and Political Ambition*. Thousand Oaks, CA: CQ.

Reckendrees, Alfred. 2015. "Weimar Germany: The First Open Access Order That Failed?" *Constitutional Political Economy* 26 (1): 38–60.

Regan, Patrick M., and M. Scott Meachum. 2014. "Data on Interventions during Periods of Political Instability." *Journal of Peace Research* 51 (1): 127–35.

Renner, Judith. 2013. *Discourse, Normative Change and the Quest for Reconciliation in Global Politics*. Manchester: Manchester University Press.

Rensmann, Lars, Sarah L. de Lange, and Stefan Couperus. 2017. "Editorial to the Issue on Populism and the Remaking of (Il)liberal Democracy in Europe." *Politics and Governance* 5 (4): 106–11.

Reus-Smit, Christian. 2012. "International Relations, Irrelevant? Don't Blame Theory." *Millennium: Journal of International Studies* 40 (3): 525–40.

Reyes, Oscar. 2005. "Skinhead Conservatism: A Failed Populist Project." In *Populism and the Mirror of Democracy*, edited by Francisco Panizza, 99–117. London: Verso.

Rich, Roland. 1993. "Recognition of States: The Collapse of Yugoslavia and the Soviet Union." *European Journal of International Law* 4 (1): 36–65.

Richardson, Laurel. 2003. "Collecting and Interpreting Qualitative Materials." In *Collecting and Interpreting Qualitative Materials*, edited by Norman K. Denzin and Yvonna S. Lincoln, 499–541. London: Sage.

Richmond, Oliver P. 2006. "The Problem of Peace: Understanding the 'Liberal Peace.'" *Conflict, Security and Development* 6 (3): 291–314.

Richmond, Oliver P. 2011. *A Post-Liberal Peace*. London: Routledge.

Risse, Thomas. 2003. "Konstruktivismus, Rationalismus und Theorien Internationaler Beziehungen—warum empirisch nichts so heiß gegessen wird, wie es theoretisch gekocht wurde." In *Die neuen Internationalen Beziehungen. Forschungsstand und Perspektiven in Deutschland*, edited by Gunther Hellmann, Klaus Dieter Wolf, and Michael Zürn, 99–132. Baden-Baden: Nomos.

Risse, Thomas. 2005. "Two-Thirds of the World: Governance in Areas of Limited Statehood Is a Global Problem." *Internationale Politik, Transatlantic Edition* 6 (4): 64–69.

Risse, Thomas. 2007. "Deutsche Identität und Außenpolitik." In *Handbuch zur deutschen Außenpolitik*, edited by Gunther Hellmann, Reinhard Wolf, and Siegmar Schmidt, 49–61. Wiesbaden: VS Verlag für Sozialwissenschaften.

Risse, Thomas. 2010. "Begrenzte Staatlichkeit und neue Governance-Strukturen." In *Einsatz für den Frieden: Sicherheit und Entwicklung in Räumen begrenzter Staatlichkeit*, edited by Josef Braml, Thomas Risse, and Eberhard Sandschneider, 23–29. Munich: R. Oldenbourg.

Risse-Kappen, Thomas. 1991. "Did 'Peace through Strength' End the Cold War? Lessons from INF." *International Security* 16 (1): 162–88.

Risse-Kappen, Thomas. 1994. "Ideas Do Not Float Freely: Transnational Coalitions, Domestic Structures, and the End of the Cold War." *International Organization* 48 (2): 185–214.

Risse-Kappen, Thomas. 1995. "Democratic Peace—Warlike Democracies?" *European Journal of International Relations* 1 (4): 491–517.

Rohlfing, Ingo. 2012. *Case Studies and Causal Inference: An Integrative Framework*. Basingstoke: Palgrave Macmillan.

Rohlfing, Ingo, and Peter Starke. 2013. "Building on Solid Ground: Robust Case Selection in Multi-Method Research." *Swiss Political Science Review* 19 (4): 492–512.

Roos, Ulrich. 2012. "Deutsche Außenpolitik nach der Vereinigung: Zwischen ernüchtertem Idealismus und realpolitischem Weltordnungsstreben." *Zeitschrift für Internationale Beziehungen* 19 (2): 7–40.

Rorty, Richard. 1979. *Philosophy and the Mirror of Nature*. Princeton, NJ: Princeton University Press.

Rorty, Richard. 1981. "Nineteenth-Century Idealism and Twentieth-Century Textualism." *Monist* 64 (2): 155–74.

Rorty, Richard. 1982. *Consequences of Pragmatism: Essays, 1972–1980*. Minneapolis: University of Minnesota Press.

Rosenau, James N. 1996. "Probing Puzzles Persistently: A Desirable but Improbable Future for IR Theory." In *International Theory: Positivism and Beyond*, edited by Ken Booth, Steve Smith, and Marysia Zalewski, 309–17. Cambridge: Cambridge University Press.

Rosert, Elvira. 2019. "Norm Emergence as Agenda Diffusion: Failure and Success in the Regulation of Cluster Munitions." *European Journal of International Relations* 25 (4): 1103–31.

Rosert, Elvira, and Sonja Schirmbeck. 2007. "Zur Erosion internationaler Normen." *Zeitschrift für Internationale Beziehungen* 14 (2): 253–87.

Rotberg, Robert I. 2002. "Failed States in a World of Terror." *Foreign Affairs* 81 (4): 127–40.

Rothe, Delf. 2015. *Securitizing Global Warming: A Climate of Complexity*. London: Routledge.

Rudolf, Peter. 2014. *Zur Ethik militärischer Gewalt*. SWP-Studie 6. Berlin: Stiftung Wissenschaft und Politik.

Rudzio, Wolfgang. 2003. *Das politische System der Bundesrepublik Deutschland*. 6th ed. Opladen: Leske und Budrich.

Rühe, Volker. 2011. "Vorwort: Sicherheitspolitik und Auslandseinsätze." In *Bewährungsproben einer Nation: Die Entsendung der Bundeswehr ins Ausland*, edited by Christoph Schwegmann, v–xv. Berlin: Duncker und Humblot.

Rumelili, Bahar. 2015. "Identity and Desecuritisation: The Pitfalls of Conflating Ontological and Physical Security." *Journal of International Relations and Development* 18 (1): 52–74.

Russett, Bruce. 2005. "Bushwhacking the Democratic Peace." *International Studies Perspectives* 6 (4): 395–408.

Sabaratnam, Meera. 2018. *Decolonising Intervention: International Statebuilding in Mozambique*. London: Rowman and Littlefield.

Said, Edward W. 1979. *Orientalism*. New York: Vintage.

Salter, Mark B. 2002. *Barbarians and Civilization in International Relations*. London: Pluto.

Salter, Mark B. 2013a. "Introduction." *Research Methods in Critical Security Studies: An Introduction*, edited by Mark B. Salter and Can E. Mutlu, 1–14. London: Routledge.

Salter, Mark B. 2013b. "Research Design: Introduction." In *Research Methods in Critical Security Studies: An Introduction*, edited by Mark B. Salter and Can E. Mutlu, 15–23. London: Routledge.

Salvatore, Jessica Di, and Andrea Ruggeri. 2017. "Effectiveness of Peacekeeping Operations." In *Oxford Research Encyclopedia of Politics*, edited by William R. Thompson. Oxford: Oxford University Press, https://doi.org/10.1093/acrefore/9780190228637.013.586.

Samuels, Richard J. 2007. *Securing Japan: Tokyo's Grand Strategy and the Future of East Asia*. Ithaca, NY: Cornell University Press.

Sanders, David. 2002. "Behaviouralism." In *Theory and Methods in Political Science*, edited by David Marsh and Gerry Stoker, 45–64. London: Palgrave Macmillan.

Sangar, Eric. 2015. "The Weight of the Past(s): The Impact of the Bundeswehr's Use of Historical Experience on Strategy-Making in Afghanistan." *Journal of Strategic Studies* 38 (4): 411–44.

Sankey, Howard. 2010. "Witchcraft, Relativism and the Problem of the Criterion." *Erkenntnis* 72 (1): 1–16.

Sarcinelli, Ulrich, and Marcus Menzel. 2007. "Medien." In *Handbuch zur deutschen Außenpolitik*, edited by Gunther Hellmann, Reinhard Wolf, and Siegmar Schmidt, 326–35. Wiesbaden: VS Verlag für Sozialwissenschaften.

Sarotte, Mary Elise. 2001. *German Military Reform and European Security*. Adelphi Papers no. 340. London: Routledge.

Sayer, Andrew. 1993. "Postmodernist Thought in Geography: A Realist View." *Antipode* 25 (4): 320–44.

Sayre-McCord, Geoff. 2012. "Metaethics." In *The Stanford Encyclopedia of Philosophy*, edited by Edward N. Zalta. Spring ed. Stanford: Stanford University, http://plato.stanford.edu/archives/spr2012/entries/metaethics/.

Schafer, Mark, and Stephen G. Walker. 2006. "Democratic Leaders and the Democratic Peace: The Operational Codes of Tony Blair and Bill Clinton." *International Studies Quarterly* 50 (3): 561–83.

Schafranek, Niels. 2009. "Deutsches Recht im Auslandseinsatz: Eine Friedensarmee auf dem Weg in die Einsatzrealität." In *Armee im Einsatz: Grundlagen, Strategien und Ergebnisse einer Beteiligung der Bundeswehr*, edited by Hans J. Gießmann and Armin Wagner, 134–47. Baden-Baden: Nomos.

Scharpf, Fritz W. 1997. *Games Real Actors Play: Actor-Centered Institutionalism in Policy Research*. Boulder, CO: Westview.

Schemo, Diana Jean. 1991. "Germany's Lukewarm Support of Gulf War Leaves Its Allies Cold." *Baltimore Sun*, 13 March 1991, http://articles.baltimoresun. com/1991-03-13/news/1991072044_1_gulf-war-persian-gulf-germany.

Schimpf, Axel. 2012. "Deutsche Marine: Klarer Kurs und klares Ziel." *Europäische Sicherheit und Technik* (October 2012): 40–44.

Schlag, Gabi, and Axel Heck. 2012. "Securitizing Images: The Female Body and the War in Afghanistan." *European Journal of International Relations* 19 (4): 891–913.

Schlichte, Klaus. 1998. *Why States Decay: A Preliminary Assessment*. Working Paper 2. Hamburg: Research Unit of Wars, Armament and Development, University of Hamburg.

Schlichte, Klaus. 2008. "'Staatszerfall' und die Dilemmata der intervenierenden Demokratie." In *Bedrohungen der Demokratie*, edited by André Brodocz, Marcus Llanque and Gary S. Schaal, 136–51. Wiesbaden: VS Verlag für Sozialwissenschaften.

Schlichte, Klaus, and Boris Wilke. 2000. "Der Staat und einige seiner Zeitgenossen: Zur Zukunft des Regierens in der 'Dritten Welt.'" *Zeitschrift für Internationale Beziehungen* 7 (2): 359–84.

Schliesky, Utz. 2014. "Die wehrhafte Demokratie des Grundgesetzes." In *Handbuch des Staatsrechts der Bundesrepublik Deutschland*, vol. 12, edited by Josef Isensee and Paul Kirchhoff, 847–77. Heidelberg: C. F. Müller Verlag.

Schlotter, Peter. 2008. "Berliner Friedenspolitik? Zum Stand der Forschung und zur Einführung." In *Berliner Friedenspolitik? Militärische Transformation—zivile Impulse—europäische Einbindung*, edited by Peter Schlotter, Wilhelm Nolte, and Renate Grasse, 7–37. Baden-Baden: Nomos.

Schmidt, Vivien A. 2008. "Discursive Institutionalism: The Explanatory Power of Ideas and Discourse." *Annual Review of Political Science* 11 (1): 303–26.

Schmidt, Vivien A. 2010. "Taking Ideas and Discourse Seriously: Explaining Change through Discursive Institutionalism as the Fourth 'New Institutionalism.'" *European Political Science Review* 2 (1): 1–25.

Schmidt, Vivien A. 2017. "Theorizing Ideas and Discourse in Political Science: Intersubjectivity, Neo-Institutionalisms, and the Power of Ideas." *Critical Review* 29 (2): 248–63.

Schmidt, Vivien A., and Claudio M. Radaelli. 2004. "Policy Change and Discourse in Europe: Conceptual and Methodological Issues." *West European Politics* 27 (2): 183–210.

Schmidtke, Henning, and Frank Nullmeier. 2011. "Political Valuation Analysis and the Legitimacy of International Organizations." *German Policy Studies* 7 (3): 117.

Schneckener, Ulrich. 2007. *International Statebuilding: Dilemmas, Strategies and Challenges for German Foreign Policy*. SWP Research Paper RP 9, Berlin: German Institute for International and Security Affairs.

Schneider, Gerald. 2017. "Capitalist Peace Theory: A Critical Appraisal." In *Oxford Research Encyclopedia of Politics*, edited by William R. Thompson. Oxford: Oxford University Press, https://doi.org/ 10.1093/ acrefore/9780190228637.013.314.

Schneider, Jörn, and Thomas Ritter. 2012. "Die Division Schnelle Kräfte." *Europäische Sicherheit und Technik* (April 2012): 34–38.

Schoen, Harald. 2010. "Ein Bericht von der Heimatfront: Bürger, Politiker und der Afghanistaneinsatz der Bundeswehr." *Politische Vierteljahresschrift* 51 (3): 395–408.

Schoenes, Katharina. 2011. "'Talibanterroristen,' freundliche Helfer und lächelnde Mädchen: Die Rolle der Frauenrechte bei der Legitimation des Afghanistan-Einsatzes der Bundeswehr." *Femina Politica* 20 (1): 78–89.

Schöllgen, Gregor. 1998. "Kriegsgefahr und Krisenmanagement vor 1914: Zur Außenpolitik des kaiserlichen Deutschland." *Historische Zeitschrift* 267 (2): 399–413.

Schooler, J. W. 2014. "Metascience Could Rescue the 'Replication Crisis.'" *Nature* 515 (7525): 9.

Schrafstetter, Susanna, and Alan E. Steinweis, eds. 2016. *The Germans and the Holocaust: Popular Responses to the Persecution and Murder of the Jews*. New York: Berghahn Books.

Schrödinger, E. 1935. "Die gegenwärtige Situation in der Quantenmechanik." *Naturwissenschaften* 23 (49): 823–28.

Schwab-Trapp, Michael. 2002. *Kriegsdiskurse: Die politische Kultur des Krieges im Wandel 1991–1999*. Opladen: Leske und Budrich.

Schwab-Trapp, Michael. 2003. "Der Nationalsozialismus im öffentlichen Diskurs über militärische Gewalt." In *Die NS-Diktatur im deutschen Erinnerungsdiskurs*, edited by Wolfgang Bergem, 171–85. Wiesbaden: VS Verlag für Sozialwissenschaften.

Schwab-Trapp, Michael. 2007. *Kampf dem Terror: Vom Anschlag auf das World Trade Center bis zum Beginn des Irakkrieges; Eine empirische Studie über die politische Kultur Deutschlands im zweiten Jahrzehnt nach der Wiedervereinigung*. Cologne: Verlag Rüdiger Köppe.

Schwab-Trapp, Michael. 2008. "Methodische Aspekte der Diskursanalyse: Probleme der Analyse diskursiver Auseinandersetzungen am Beispiel der deutschen Diskussion über den Kosovokrieg." In *Handbuch Sozialwissenschaftliche Diskursanalyse, vol. 2, Forschungspraxis*, 3rd ed., edited by Andreas Hirseland, Reiner Keller, Werner Schneider and Willy Viehöver, 171–96. Opladen: Leske und Budrich.

Schwartz-Shea, Peregrine, and Dvora Yanow. 2012. *Interpretive Research Design: Concepts and Processes*. London: Routledge.

Seale, Clive. 1999. "Quality in Qualitative Research." *Qualitative Inquiry* 5 (4): 465–78.

Seawright, J., and J. Gerring. 2008. "Case Selection Techniques in Case Study Research: A Menu of Qualitative and Quantitative Options." *Political Research Quarterly* 61 (2): 294–308.

Shannon, Vaughn. 2017. "International Norms and Foreign Policy." In *Oxford Research Encyclopedia of Politics*, edited by William R. Thompson. Oxford: Oxford University Press, https://doi.org/ 10.1093/ acrefore/9780190228637.013.442.

Shepherd, Laura J. 2006. "Veiled References: Constructions of Gender in the Bush Administration Discourse on the Attacks on Afghanistan Post-9/11." *International Feminist Journal of Politics* 8 (1): 19–41.

Shepherd, Laura J. 2007. "'Victims, Perpetrators and Actors' Revisited: Exploring the Potential for a Feminist Reconceptualisation of (International) Security and (Gender) Violence." *British Journal of Politics and International Relations* 9 (2): 239–56.

Shim, David. 2014. *Visual Politics and North Korea: Seeing Is Believing*. London: Routledge.

Shim, David, and Frank A. Stengel. 2017. "Social Media, Gender and the Mediatisation of War: Exploring the German Armed Forces' Visual Representation of the Afghanistan Operation on Facebook." *Global Discourse* 7 (2–3): 330–47.

Shortland, N., H. Sari, and E. Nader. 2019. "Recounting the Dead: An Analysis of ISAF Caused Civilian Casualties in Afghanistan." *Armed Forces and Society* 45 (1): 122–39.

Siegelberg, Jens. 1994. *Kapitalismus und Krieg: Eine Theorie des Krieges in der Weltgesellschaft.* Münster: Lit.

Silberzahn, Raphael, and Eric L. Uhlmann. 2015. "Crowdsourced Research: Many Hands Make Tight Work." *Nature* 526 (7572): 189–91.

Simma, Bruno. 1999. "NATO, the UN and the Use of Force: Legal Aspects." *European Journal of International Law* 10 (1): 1–22.

Singer, Peter. 1993. *Practical Ethics.* 2nd ed. Cambridge: Cambridge University Press.

Siwert-Probst, Judith. 1993. "Die klassischen außenpolitischen Institutionen." In *Deutschlands neue Außenpolitik, vol. 4, Institutionen und Ressourcen,* edited by Wolf-Dieter Eberwein and Karl Kaiser, 13. Munich: R. Oldenbourg.

Sjoberg, Laura, ed. 2010. *Gender and International Security: Feminist Perspectives.* New York: Routledge.

Sjoberg, Laura. 2011. "Gender, the State, and War Redux: Feminist International Relations across the 'Levels of Analysis.'" *International Relations* 25 (1): 108–34.

Sjoberg, Laura, and J. Ann Tickner. 2013. "Feminist Perspectives on International Relations." In *Handbook of International Relations,* edited by Walter Carlsnaes, Thomas Risse, and Beth A. Simmons, 170–95. 2nd ed. London: Sage.

Sjöstedt, Roxanna. 2013. "Ideas, Identities and Internalization: Explaining Securitizing Moves." *Cooperation and Conflict* 48 (1): 143–64.

Skinner, Quentin. 2002. *Visions of Politics. Vol. 1, Regarding Method.* Cambridge: Cambridge University Press.

Sloan, James. 2011. *The Militarisation of Peacekeeping in the Twenty-First Century.* Oxford: Hart.

Smith, Linda Tuhiwai. 2013. *Decolonizing Methodologies: Research and Indigenous Peoples.* London: Zed Books.

Smith, Steve. 2004. "Singing Our World into Existence: International Relations Theory and September 11." *International Studies Quarterly* 48 (3): 499–515.

Solomon, Thomas. 2012. "Theory and Method in Popular Music Analysis: Text and Meaning." *Studia Musicologica Norvegica* 38 (5): 86–108.

Solomon, Ty. 2014. *The Politics of Subjectivity in American Foreign Policy Discourses.* Ann Arbor: University of Michigan Press.

Sontheimer, Kurt, and Wilhelm Bleek. 2002. *Grundzüge des politischen Systems Deutschlands.* Bonn: Bundeszentrale für politische Bildung.

Spencer, Alexander. 2014. "Romantic Stories of the Pirate in IARRRH: The Failure of Linking Piracy and Terrorism Narratives in Germany." *International Studies Perspectives* 15 (3): 297–312.

Spitzmüller, Jürgen, and Ingo H. Warnke. 2011. "Discourse as a 'Linguistic Object': Methodical and Methodological Delimitations." *Critical Discourse Studies* 8 (2): 75–94.

Spivak, Gayatri Chakravorty. 1985. "Three Women's Texts and a Critique of Imperialism." *Critical Inquiry* 12 (1): 243–61.

Spivak, Gayatri Chakravorty. 1988. "Can the Subaltern Speak?" In *Marxism and the Interpretation of Culture*, edited by Cary Nelson and Lawrence Grossberg, 271–316. Champaign: University of Illinois Press.

Sprenger, Sebastian. 2019. "More Money, More Missions: German Defense Minister Unveils Her Plan for the Bundeswehr." *Defense News*, 7 November, https://www.defensenews.com/global/europe/2019/11/07/more-money-more-missions-german-defense-minister-unveils-her-plan-for-the-bundeswehr/.

Springer, Simon, Kean Birch, and Julie MacLeavy, eds. 2017. *The Handbook of Neoliberalism*. London: Routledge.

Stachowitsch, Saskia. 2013. "Professional Soldier, Weak Victim, Patriotic Heroine." *International Feminist Journal of Politics* 15 (2): 157–76.

Stäheli, Urs. 2006. "Die politische Theorie der Hegemonie: Ernesto Laclau und Chantal Mouffe." In *Politische Theorien der Gegenwart II: Eine Einführung*, edited by André Brodocz and Gary S. Schaal, 253–84. Opladen: Leske und Budrich.

Stahl, Bernhard. 2017. "Verantwortung—welche Verantwortung? Der deutsche Verantwortungsdiskurs und die Waffenlieferungen an die Peschmerga." *Zeitschrift für Politikwissenschaft* 27 (4): 437–71.

Stahl, Bernhard, Robin Lucke, and Anna Felfeli. 2016. "Comeback of the Transatlantic Security Community? Comparative Securitisation in the Crimea Crisis." *East European Politics* 32 (4): 525–46.

Stark Urrestarazu, Ursula. 2015. "Neue Macht, neue Verantwortung, neue Identität? 'Deutschlands Rolle in der Welt' aus identitätstheoretischer Perspektive." *Zeitschrift für Außen- und Sicherheitspolitik* 8 (1): 173–95.

Stavrianakis, Anna, and Maria Stern. 2018. "Militarism and Security: Dialogue, Possibilities and Limits." *Security Dialogue* 49 (1–2): 3–18.

Steans, Jill. 2003. "Engaging from the Margins: Feminist Encounters with the 'Mainstream' of International Relations." *British Journal of Politics and International Relations* 5 (3): 428–54.

Steele, Brent J. 2008. *Ontological Security in International Relations: Self-Identity and the IR State*. London: Routledge.

Steele, Brent J. 2019. *Restraint in International Politics*. Cambridge: Cambridge University Press.

Steinmeier, Frank-Walter. 2016. "Germany's New Global Role." *Foreign Affairs* 95 (4): 106–13.

Stengel, Frank A. 2007. "Taking Stock: The Focal Points of Abe's Foreign Policy." *Japan aktuell: Journal of Current Japanese Affairs* 15 (6): 53–71.

Stengel, Frank A. 2008. "The Reluctant Peacekeeper: Japan's Ambivalent Stance on UN Peace Operations." *Japan aktuell: Journal of Current Japanese Affairs* 16 (1): 37–55.

Stengel, Frank A. 2010. "Légitimer l'armée en operation: Les interventions extérieures de la 'nouvelle Bundeswehr' dans la rhétorique du gouvernement rouge-vert." *Allemagne d'aujourd'hui* (192): 25–34.

Stengel, Frank A. 2019a. "The Political Production of Ethical War: Rethinking the Ethics/Politics Nexus with Laclau." *Critical Studies on Security* 7 (3): 230–42.

Stengel, Frank A. 2019b. "Securitization as Discursive (Re)articulation: Explaining the Relative Effectiveness of Threat Construction." *New Political Science* 41 (2): 294–312.

Stengel, Frank A., and Rainer Baumann. 2018. "Non-state Actors and Foreign Policy." In *The Oxford Encyclopedia of Foreign Policy Analysis*, edited by Cameron Thies, 266–86. Oxford: Oxford University Press.

Stengel, Frank A., and Dirk Nabers. 2019. "Symposium: The Contribution of Laclau's Discourse Theory to International Relations and International Political Economy: Introduction." *New Political Science* 41 (2): 248–62.

Stengel, Frank A., and Christoph Weller. 2008. *Vier Jahre Aktionsplan "Zivile Krisenprävention"—war das alles?* GIGA Focus Global 11. Hamburg: GIGA German Institute of Global and Area Studies.

Stengel, Frank A., and Christoph Weller. 2010. "Action Plan or Faction Plan? Germany's Eclectic Approach to Conflict Prevention." *International Peacekeeping* 17 (1): 93–107.

Stern, Maria. 2011. "Gender and Race in the European Security Strategy: Europe as a 'Force for Good'?" *Journal of International Relations and Development* 14 (1): 28–59.

Stern, Maria, and Joakim Öjendal. 2010. "Mapping the Security-Development Nexus: Conflict, Complexity, Cacophony, Convergence?" *Security Dialogue* 41 (1): 5–29.

Steup, Matthias. 2014. "Epistemology." In *The Stanford Encyclopedia of Philosophy*, edited by Edward N. Zalta. Spring ed. Stanford: Stanford University, https://plato.stanford.edu/entries/epistemology/.

Stocking, George W. 1968. "Empathy and Antipathy in the Heart of Darkness: An Essay Review of Malinowski's Field Diaries." *Journal of the History of the Behavioral Sciences* 4 (2): 189–94.

Stöhs, Jeremy. 2018. "Into the Abyss: European Naval Power in the Post-Cold War Era." *Naval War College Review* 71 (3): article 4.

Stritzel, Holger. 2012. "Securitization, Power, Intertextuality: Discourse Theory and the Translations of Organized Crime." *Security Dialogue* 43 (6): 549–67.

Stritzel, Holger. 2014. *Security in Translation: Securitization Theory and the Localization of Threat*. Basingstoke: Palgrave Macmillan.

Subotić, Jelena. 2016. "Narrative, Ontological Security, and Foreign Policy Change." *Foreign Policy Analysis* 12 (4): 610–27.

Suzuki, Shogo. 2007. "The Importance of 'Othering' in China's National Identity: Sino-Japanese Relations as a Stage of Identity Conflicts." *Pacific Review* 20 (1): 23–47.

Sweetman, Brendan. 1999. "Postmodernism, Derrida and Différance: A Critique." *International Philosophical Quarterly* 39 (1): 5–18.

Swoyer, Chris. 2010. "Relativism." In *The Stanford Encyclopedia of Philosophy*, edited by Edward N. Zalta. Winter ed. Stanford: Stanford University, http://plato.stanford.edu/archives/win2010/entries/relativism/.

Sylvan, Donald A, ed. 1998. *Problem Representation in Foreign Policy Decision Making*. Cambridge: Cambridge University Press.

Sylvan, Donald A., Andrea Grove, and Jeffrey D. Martinson. 2005. "Problem Representation and Conflict Dynamics in the Middle East and Northern Ireland." *Foreign Policy Analysis* 1 (3): 279–99.

Tallis, Benjamin. 2020. "An International Politics of Czech Architecture; or, Reviving the International in International Political Sociology." *Globalizations* 17 (3): 452–76.

Tannenwald, Nina. 1999. "The Nuclear Taboo: The United States and Normative Basis of Nuclear Non-use." *International Organization* 53 (3): 433–68.

Tetlock, Philip E. 1998. "Close-Call Counterfactuals and Belief System Defense: I Was Not Almost Wrong but I Was Almost Right." *Journal of Personality and Social Psychology* 75: 230–42.

Thomas, Claire. 2011. "Why Don't We Talk about 'Violence' in International Relations?" *Review of International Studies* 37 (4): 1815–36.

Thomassen, Lasse. 2005. "Antagonism, Hegemony and Ideology after Heterogeneity." *Journal of Political Ideologies* 10 (3): 289–309.

Thränhardt, Dietrich. 2000. "Bundesregierung." In *Handwörterbuch des politischen Systems der Bundesrepublik Deutschland*, edited by Uwe Andersen and Wichard Woyke, 60–66. Bonn: Bundeszentrale für politische Bildung.

Tickner, J. Ann. 1988. "Hans Morgenthau's Principles of Political Realism: A Feminist Reformulation." *Millennium: Journal of International Studies* 17 (3): 429–40.

Tickner, J. Ann. 1992. *Gender in International Relations: Feminist Perspectives on Achieving Global Security*. New York: Columbia University Press.

Tickner, J. Ann. 1997. "You Just Don't Understand: Troubled Engagements between Feminists and IR Theorists." *International Studies Quarterly* 41 (4): 611–32.

Tilly, Charles. 2003. *The Politics of Collective Violence*. Cambridge: Cambridge University Press.

Tlostanova, Madina. 2012. "Postsocialist ≠ Postcolonial? On Post-Soviet Imaginary and Global Coloniality." *Journal of Postcolonial Writing* 48 (2): 130–42.

Torfing, Jacob. 1999. *New Theories of Discourse: Laclau, Mouffe and Žižek*. Oxford: Blackwell.

Torfing, Jacob. 2005a. "Discourse Theory: Achievements, Arguments, and Challenges." In *Discourse Theory in European Politics*, edited by David Howarth and Jacob Torfing, 1–32. Basingstoke: Palgrave Macmillan.

Torfing, Jacob. 2005b. "Poststructuralist Discourse Theory: Foucault, Laclau, Mouffe, and Žižek." In *The Handbook of Political Sociology: States, Civil Societies, and Globalization*, edited by Thomas Janoski, Robert R. Alford, Alexander M. Hicks, and Mildred A. Schwartz, 153–71. Cambridge: Cambridge University Press.

Towns, Ann. 2019. "Gender Troubled? Three Simple Steps to Avoid Silencing Gender in IR." *E-International Relations*, 15 February, https://www.e-ir.info/2019/02/15/gender-troubled-three-simple-steps-to-avoid-silencing-gender-in-ir/.

Tracy, S. J. 2010. "Qualitative Quality: Eight 'Big-Tent' Criteria for Excellent Qualitative Research." *Qualitative Inquiry* 16 (10): 837–51.

Troche, Alexander. 2000. "'Ich habe nur die Hoffnung, dass der Kelch an uns vorübergeht . . .': Der Zypernkonflikt und die erste deutsche Out-of-area-Entscheidung." *Historisch-politische Mitteilungen* 7: 183–95.

True, Jacqui, and Sarah Hewitt. 2018. "International Relations and the Gendered International." In *The Sage Handbook of the History, Philosophy and Sociology of International Relations*, edited by Andreas Gofas, Inanna Hamati-Ataya, and Nicholas Onuf, 90–104. London: Sage.

Ulbert, Cornelia. 2012. "Vom Klang vieler Stimmen: Herausforderungen 'kritischer' Normenforschung: Eine Replik auf Stephan Engelkamp, Katharina Glaab und Judith Renner." *Zeitschrift für Internationale Beziehungen* 19 (2): 129–39.

UNAMA. 2018. *Afghanistan: Protection of Civilians in Armed Conflict; Annual Report 2017*. Kabul: United Nations Assistance Mission in Afghanistan.

van Dijk, Teun A. 1997, ed. *Discourse as Structure and Process*. London: Sage.

van Dijk, Teun A. 2006a. "Discourse, Context and Cognition." *Discourse Studies* 8 (1): 159–77.

van Dijk, Teun A. 2006b. "Ideology and Discourse Analysis." *Journal of Political Ideologies* 11 (2): 115–40.

Van Evera, Stephen. 1997. *Guide to Methods for Students of Political Science*. Ithaca, NY: Cornell University Press.

van Inwagen, Peter, and Meghan Sullivan. 2015. "Metaphysics." In *The Stanford Encyclopedia of Philosophy*, edited by Edward N. Zalta. Spring ed. Stanford: Stanford University, http://plato.stanford.edu/entries/metaphysics/.

Vázquez, Rolando. 2011. "Translation as Erasure: Thoughts on Modernity's Epistemic Violence." *Journal of Historical Sociology* 24 (1): 27–44.

Viehoff, Daniel. 2014. "Democratic Equality and Political Authority." *Philosophy and Public Affairs* 42 (4): 337–75.

von Alemann, Ulrich. 2000. *Das Parteiensystem der Bundesrepublik Deutschland*. Bonn: Bundeszentrale für politische Bildung.

von Bredow, Wilfried. 2015. *Sicherheit, Sicherheitspolitik und Militär: Deutschland seit der Vereinigung*. Wiesbaden: Springer VS.

von Krause, Ulf. 2013. *Die Bundeswehr als Instrument deutscher Außenpolitik*. Wiesbaden: Springer VS.

von Krause, Ulf. 2015. *Bundeswehr und Außenpolitik: Zur Rolle des Militärs im Diskurs um mehr Verantwortung Deutschlands in der Welt*. Wiesbaden: Springer VS.

von Weizsäcker, Richard. 2000. *Gemeinsame Sicherheit und Zukunft der Bundeswehr: Bericht der Kommission an die Bundesregierung*. Bonn: Kommission Gemeinsame Sicherheit und Zukunft der Bundeswehr.

Vorländer, Hans. 2006. "Deutungsmacht—Die Macht der Verfassungsgerichts-barkeit." In *Die Deutungsmacht der Verfassungsgerichtsbarkeit*, edited by Hans Vorländer, 9–33. Wiesbaden: VS Verlag für Sozialwissenschaften.

Vucetic, Srdjan. 2017. "Identity and Foreign Policy." In *Oxford Research Encyclopedia of Politics*, edited by William R. Thompson. Oxford: Oxford University Press, https://doi.org/ 10.1093/acrefore/9780190228637.013.435.

Vucetic, Srdjan, and Randolph B. Persaud. 2018. "Race and International Relations." In *Race, Gender, and Culture in International Relations: Postcolonial Perspectives*, edited by Randolph B. Persaud and Alina Sajed, 35–57. London: Routledge.

Wæver, Ole. 2004. "Discursive Aproaches." In *European Integration Theory*, edited by Antje Wiener and Thomas Diez, 195–215. Oxford: Oxford University Press.

Wagenaar, Hendrik. 2011. *Meaning in Action: Interpretation and Dialogue in Policy Analysis*. Armonk, NY: M. E. Sharpe.

Wagner, Wolfgang. 2005. "Die soziale Konstruktion außenpolitischer Interessen: Deutsche und britische Debatten über eine Stärkung der Gemeinsamen Außen- und Sicherheitspolitik der Europäischen Union." In *Konstruktivistische Analysen der Internationalen Politik*, edited by Cornelia Ulbert and Christoph Weller, 65–97. Wiesbaden: VS Verlag für Sozialwissenschaften.

Wagner, Wolfgang, Anna Herranz-Surrallés, Juliet Kaarbo, and Falk Ostermann. 2018. "Party Politics at the Water's Edge: Contestation of Military Operations in Europe." *European Political Science Review* 10 (4): 537–63.

Wajner, Daniel F. 2019. "'Battling' for Legitimacy: Analyzing Performative Contests in the Gaza Flotilla Paradigmatic Case." *International Studies Quarterly* 63 (4): 1035–50.

Walker, R. B. J. 1993. *Inside/Outside: International Relations as Political Theory*. Cambridge: Cambridge University Press.

Waltz, Kenneth N. 1979. *Theory of International Politics*. Boston: McGraw-Hill.

Weber, Cynthia. 1999. *Faking It: US Hegemony in a "Post-Phallic" Era*. Minneapolis: University of Minnesota Press.

Weber, Cynthia. 2014a. "From Queer to Queer IR." *International Studies Review* 16 (4): 596–601.

Weber, Cynthia. 2014b. "Why Is There No Queer International Theory?" *European Journal of International Relations* 21 (1): 27–51.

Wegener, Jens. 2007. "Shaping Germany's Post-War Intelligence Service: The Gehlen Organization, the U.S. Army, and Central Intelligence, 1945–1949." *Journal of Intelligence History* 7 (1): 41–59.

Wehner, Leslie E., and Cameron G. Thies. 2014. "Role Theory, Narratives, and Interpretation: The Domestic Contestation of Roles." *International Studies Review* 16 (3): 411–36.

Weise, Frank-Jürgen, Hans Heinrich Driftmann, Hans-Ulrich Klose, Jürgen Kluge, Karl-Heinz Lather, and Helga von Wedel. 2010. *Vom Einsatz her Denken: Konzentration, Flexibilität, Effizienz; Bericht der Strukturkommission der Bundeswehr*. Berlin: Federal Ministry of Defence.

Welch, David A. 2005. *Painful Choices: A Theory of Foreign Policy Change*. Princeton, NJ: Princeton University Press.

Weldes, Jutta. 1999. *Constructing National Interests: The United States and the Cuban Missile Crisis*. Minneapolis: University of Minnesota Press.

Weldes, Jutta, Mark Laffey, Hugh Gusterson, and Raymond Duvall, eds. 1999. *Cultures of Insecurity: States, Communities, and the Production of Danger*. Minneapolis: University of Minnesota Press.

Weldes, Jutta, and Diana Saco. 1996. "Making State Action Possible: The United States and the Discursive Construction of the Cuban Problem, 1960–1994." *Millennium: Journal of International Studies* 25 (2): 361–95.

Welland, Julia. 2015. "Liberal Warriors and the Violent Colonial Logics of 'Partnering and Advising.'" *International Feminist Journal of Politics* 17 (2): 289–307.

Weller, Christoph. 2005. "Perspektiven eines reflexiven Konstruktivismus für die Internationalen Beziehungen." In *Konstruktivistische Analysen der internationalen Politik*, edited by Cornelia Ulbert and Christoph Weller, 35–64. Wiesbaden: VS Verlag für Sozialwissenschaften.

Weller, Christoph. 2007. "Bundesministerien." In *Handbuch zur deutschen Außenpolitik*, edited by Gunther Hellmann, Reinhard Wolf, and Siegmar Schmidt, 210–23. Wiesbaden: VS Verlag für Sozialwissenschaften.

Weller, Christoph. 2008. "Zivile Krisenprävention und Konfliktbearbeitung: Politische Herausforderungen und der Aktionsplan der Bundesregierung." In *Berliner Friedenspolitik? Militärische Transformation—zivile Impulse—europäische Einbindung*, edited by Peter Schlotter, Wilhelm Nolte, and Renate Grasse, 109–36. Baden-Baden: Nomos.

Weller, Christoph, and Andrea Kirschner. 2005. "Zivile Konfliktbearbeitung—Allheilmittel oder Leerformel? Möglichkeiten und Grenzen eines viel versprechenden Konzepts." *Internationale Politik und Gesellschaft* (4): 10–29.

Weller, Marc. 1992. "The International Response to the Dissolution of the Socialist Federal Republic of Yugoslavia." *American Journal of International Law* 86 (3): 569–607.

Wendt, Alexander. 1987. "The Agent-Structure Problem in International Relations." *International Organization* 41 (3): 335–70.

Wendt, Alexander. 1991. "Bridging the Theory/Meta-Theory Gap in International Relations." *Review of International Studies* 17 (4): 383–92.

Wendt, Alexander. 1995. "Constructing International Politics." *International Security* 20 (1): 71–81.

Wendt, Alexander. 1999. *Social Theory of International Politics*. Cambridge: Cambridge University Press.

Wendt, Alexander. 2006. "Social Theory as Cartesian Science: An Auto-Critique from a Quantum Perspective." In *Constructivism and International Relations: Alexander Wendt and His Critics*, edited by Stefano Guzzini and Anna Leander, 181–219. London: Routledge.

Wendt, Alexander. 2015. *Quantum Mind and Social Science: Unifying Physical and Social Ontology*. Cambridge: Cambridge University Press.

Wendt, Alexander, and Raymond Duvall. 2008. "Sovereignty and the UFO." *Political Theory* 36 (4): 607–33.

Western, Jon. 2005. "The War over Iraq: Selling the War to the American Public." *Security Studies* 14 (1): 106–39.

White, Jay D. 1992. "Taking Language Seriously: Toward a Narrative Theory of Knowledge for Administrative Research." *American Review of Public Administration* 22 (2): 75–88.

Wibben, Annick T. R. 2018. "Why We Need to Study (US) Militarism: A Critical Feminist Lens." *Security Dialogue* 49 (1–2): 136–48.

Wibben, Annick T. R. 2020. "Everyday Security, Feminism, and the Continuum of Violence." *Journal of Global Security Studies* 5 (1):115–21.

Widmaier, Wesley W., Mark Blyth, and Leonard Seabrooke. 2007. "Exogeneous Shocks or Endogeneous Constructions? The Meanings of Wars and Crises." *International Studies Quarterly* 51 (4): 747–59.

Wiedemann, Gregor. 2013. "Opening Up to Big Data: Computer-Assisted Analysis of Textual Data in Social Sciences." *Forum Qualitative Sozialforschung / Forum: Qualitative Social Research* 14 (2), https://doi.org/10.17169/fqs-14.2.1949.

Wiefelspütz, Dieter. 2003. "Der Einsatz der Streitkräfte und die konstitutive Beteiligung des deutschen Bundestages." *Neue Zeitschrift für Wehrrecht* 45 (4): 133–51.

Wiefelspütz, Dieter. 2009. "Der Auslandseinsatz der Streitkräfte und der Deutsche Bundestag." In *Müssen Parlamentsreformen scheitern?*, edited by Julia Blumenthal and Stephan Bröchler, 109–46. Wiesbaden: VS Verlag für Sozialwissenschaften.

Wiener, Antje. 2009. "Enacting Meaning-in-Use: Qualitative Research on Norms and International Relations." *Review of International Studies* 35 (1): 175–93.

Wiener, Antje. 2014. *A Theory of Contestation*. Berlin: Springer.

Wierzbicka, Anna. 2013a. *Imprisoned in English: The Hazards of English as a Default Language*. Oxford: Oxford University Press.

Wierzbicka, Anna. 2013b. "Translatability and the Scripting of Other Peoples' Souls." *Australian Journal of Anthropology* 24 (1): 1–21.

Wiesner, Ina. 2011. "Die Transformation der Bundeswehr in Deutschland." In *Transformation der Sicherheitspolitik*, edited by Thomas Jäger and Ralph Thiele, 91–104. Wiesbaden: VS Verlag für Sozialwissenschaften.

Wight, Colin. 1999a. "MetaCampbell: The Epistemological Problems of Perspectivism." *Review of International Studies* 25 (2): 311–16.

Wight, Colin. 1999b. "They Shoot Dead Horses Don't They? Locating Agency in the Agent-Structure Problematique." *European Journal of International Relations* 5 (1): 109–42.

Wight, Colin. 2007a. "Inside the Epistemological Cave All Bets Are Off." *Journal of International Relations and Development* 10 (1): 40–56.

Wight, Colin. 2007b. "A Manifesto for Scientific Realism in IR: Assuming the Can-Opener Won't Work!" *Millennium: Journal of International Studies* 35 (2): 379–98.

Wilcox, Lauren. 2014. "Queer Theory and the 'Proper Objects' of International Relations." *International Studies Review* 16 (4): 612–15.

Winch, Peter. 1990. *The Idea of a Social Science: And Its Relation to Philosophy*. 2nd ed. London: Routledge.

Wittgenstein, Ludwig. 1999. *Philosophical Investigations*. 2nd ed. Oxford: Blackwell.

Wodak, Ruth, and Michael Meyer, eds. 2001. *Methods of Critical Discourse Analysis*. London: Sage.

Wodrig, Stefanie. 2017. *Regional Intervention Politics in Africa: Crisis, Hegemony, and the Transformation of Subjectivity*. London: Routledge.

Wojczewski, Thorsten. 2018. *India's Foreign Policy Discourse and Its Conceptions of World Order: The Quest for Power and Identity*. London: Routledge.

Wolff, Jonathan. 2011. "Karl Marx." In *The Stanford Encyclopedia of Philosophy*, edited by Edward N. Zalta. Summer ed. Stanford: Stanford University, http://plato.stanford.edu/archives/sum2011/entries/marx/.

Wonka, Arndt. 2007. "Concept Specification in Political Science Research." In *Research Design in Political Science: How to Practice What They Preach?*, edited by Thomas Gschwend and Frank Schimmelpfennig, 41–61. Basingstoke: Palgrave Macmillan.

Worth, Owen. 2011. "Recasting Gramsci in International Politics." *Review of International Studies* 37 (1): 373–92.

Wrana, Daniel. 2014. "Arena." In *DiskursNetz: Wörterbuch der interdisziplinären Diskursforschung*, edited by Daniel Wrana, Alexander Ziem, Martin Reisigl, Martin Nonhoff, and Johannes Angermüller, 36–37. Frankfurt am Main: Suhrkamp.

Wullweber, Joscha. 2012. "Hegemoniale Strategien: Das Ringen um Akzeptanz in der politischen Governance der Nanotechnologie." *Leviathan* 40 (1): 4–23.

Wullweber, Joscha. 2014. "Global Politics and Empty Signifiers: The Political Construction of High Technology." *Critical Policy Studies* 9 (1): 78–96.

Yanow, Dvora. 2006. "Thinking Interpretively: Philosophical Presuppositions and the Human Sciences." In *Interpretation and Method: Empirical Research Methods and the Interpretive Turn*, edited by Dvora Yanow and Peregrine Schwartz-Shea, 5–26. New York: M. E. Sharpe.

Yanow, Dvora. 2009. "Interpretive Ways of Knowing in the Study of Politics." In *Methoden der vergleichenden Politik- und Sozialwissenschaft: Neue Entwicklungen und Anwendungen*, edited by Susanne Pickel, Gert Pickel, Hans-Joachim Lauth, and Detlef Jahn, 429–39. Wiesbaden: VS Verlag für Sozialwissenschaften.

Yanow, Dvora, and Peregrine Schwartz-Shea, eds. 2006a. *Interpretation and Method: Empirical Research Methods and the Interpretive Turn*. New York: M. E. Sharpe.

Yanow, Dvora, and Peregrine Schwartz-Shea. 2006b. "Introduction" *Interpretation and Method: Empirical Research Methods and the Interpretive Turn*, edited by Dvora Yanow and Peregrine Schwartz-Shea, xi–xxvii. New York: M. E. Sharpe.

Yennie Lindgren, Wrenn, and Petter Y. Lindgren. 2017. "Identity Politics and the East China Sea: China as Japan's 'Other.'" *Asian Politics and Policy* 9 (3): 378–401.

Young, Iris Marion. 2003. "The Logic of Masculinist Protection: Reflections on the Current Security State." *Signs: Journal of Women in Culture and Society* 29 (1): 1–25.

Young, Thomas-Durell. 1996. "German National Command Structures after Unification: A New German General Staff?" *Armed Forces and Society* 22 (3): 379–400.

Ypi, Lea. 2013. "What's Wrong with Colonialism." *Philosophy and Public Affairs* 41 (2): 158–91.

Zalewski, Marysia. 2019. "Forget(ting) Feminism? Investigating Relationality in International Relations." *Cambridge Review of International Affairs* 32 (5): 615–35.

Zalewski, Marysia, and Jane L. Parpart, eds. 2008. *Rethinking the Man Question: Sex, Gender and Violence in International Relations*. London: Zed Books.

Zehfuss, Maja. 2002. *Constructivism in International Relations: The Politics of Reality*. Cambridge: Cambridge University Press.

Zehfuss, Maja. 2007. *Wounds of Memory: The Politics of War in Germany*. Cambridge: Cambridge University Press.

Zehfuss, Maja. 2013. "Critical Theory, Poststructuralism, and Postcolonialism." In *Handbook of International Relations*, edited by Walter Carlsnaes, Thomas Risse, and Beth A. Simmons, 145–70. 2nd ed. London: Sage.

Zehfuss, Maja. 2018. *War and the Politics of Ethics*. Oxford: Oxford University Press.

Zevnik, Andreja. 2016. *Lacan, Deleuze and World Politics: Rethinking the Ontology of the Political Subject*. London: Routledge.

Ziai, Aram. 2006. "Post-Development: Ideologiekritik in der Entwicklungstheorie." *Politische Vierteljahresschrift* 47 (2): 193–218.

Ziai, Aram. 2010. "German Development Policy, 1998–2005: The Limits of Normative Global Governance." *Journal of International Relations and Development* 13 (2): 136–62.

Ziai, Aram. 2014. "Progressing Towards Incoherence: Development Discourse since the 1980s." *Momentum Quarterly* 3 (1): 3–14.

Zimmermann, Hubert. 2017. "Exporting Security: Success and Failure in the Securitization and Desecuritization of Foreign Military Interventions." *Journal of Intervention and Statebuilding* 11 (2): 225–44.

Zimmermann, Lisbeth, and Nicole Deitelhoff. 2019. "Norms under Challenge: Unpacking the Dynamics of Norm Robustness." *Journal of Global Security Studies* 4 (1): 2–17.

zu Guttenberg, Karl-Theodor. 2009. "Ich verstehe jeden, der sagt, in Afghanistan ist Krieg." *Bild Online*, 13 November 2009, http://www.bild.de/politik/2009/interview/interview-mit-minister-guttenberg-10319932.bild.html.

Zürn, Michael. 2013. "Globalization and Global Governance." In *Handbook of*

International Relations, edited by Walter Carlsnaes, Thomas Risse, and Beth A. Simmons, 401–25. end ed. London: Sage.

Zürn, Michael. 2014. "The Politicization of World Politics and Its Effects: Eight Propositions." *European Political Science Review* 6 (1): 47–71.

Zürn, Michael, Martin Binder, and Matthias Ecker-Ehrhardt. 2012. "International Authority and Its Politicization." *International Theory* 4 (1): 69–106.

Index